Geophysical and Climate Hazards: A Very Short Introduction

VERY SHORT INTRODUCTIONS are for anyone wanting a stimulating and accessible way into a new subject. They are written by experts, and have been translated into more than 45 different languages.

The series began in 1995, and now covers a wide variety of topics in every discipline. The VSI library currently contains over 750 volumes—a Very Short Introduction to everything from Psychology and Philosophy of Science to American History and Relativity—and continues to grow in every subject area.

Very Short Introductions available now:

Available soon:

For more information visit our website

www.oup.com/vsi/

Bill McGuire

GEOPHYSICAL AND CLIMATE HAZARDS

A Very Short Introduction

THIRD EDITION

OXFORD
UNIVERSITY PRESS

Great Clarendon Street, Oxford, OX2 6DP,
United Kingdom

Oxford University Press is a department of the University of Oxford.
It furthers the University's objective of excellence in research, scholarship,
and education by publishing worldwide. Oxford is a registered trade mark of
Oxford University Press in the UK and in certain other countries

© Bill McGuire 2002, 2005, 2014, 2024

The moral rights of the author have been asserted

First published in hardback as A Guide to the End of the World 2002
First Published as a Very Short Introduction as Global Catastrophes in 2002
Second edition published 2014

Published in the United States of America by Oxford University Press
198 Madison Avenue, New York, NY 10016, United States of America

British Library Cataloguing in Publication Data
Data available

Library of Congress Control Number: 2023943388

ISBN 978-0-19-287453-5

Printed and bound by
Printed by Integrated Books International, United States of America

For the millions affected by geophysical hazards every year and the billions who will reap the global heating whirlwind

Contents

Preface

In these tempestuous times, it seems that not a day goes by without news of more geophysical mayhem splashed across our Twitter feeds, television screens, and front pages. Earthquakes and volcanic eruptions, floods and storms, and other deadly phenomena, have always been with us and have taken a toll on communities for many thousands of years. But now things are changing—and fast. Global heating, unequivocally a consequence of humankind's carbon-polluting activities, is building rapidly, and in recent years has translated faster than anyone could have imagined into ever more extreme weather. As a consequence, many of the phenomena that we used to call *natural* hazards, can no longer be thought of as natural at all. Instead, to an increasing degree, the hand of humankind can be detected almost anywhere that disaster rears its head. For this reason, the term *geophysical*—rather than natural—hazard is used throughout this book, with those exacerbated by global heating forming part of a climate hazard subset.

Unsurprisingly, in a hotter world, heatwaves, wildfires, and droughts have, in the past few years, become supercharged. Temperature records have been shattered across the planet, and continue to be broken year on year. Wildfires have ravaged millions of acres of forest, and obliterated or severely damaged communities in places as far apart as Canada, California,

Australia, Spain, Turkey, Russia, and, extraordinarily, even the UK.

An overheating planet also means more intense rainfall, and as a consequence increasingly severe flooding. In the last few years alone, unprecedented floods have afflicted every continent, culminating—in July 2022—in floodwaters in Pakistan destroying or damaging more than two million homes and displacing 33 million people. This is part of a trend, building on devastating flooding in 2021 in parts of Europe, eastern Australia, China, Turkey, and elsewhere, of which no end is in sight.

Storms are becoming increasingly energized too, as the oceans from which many draw strength become ever hotter. Tropical cyclones are already more powerful, wetter, and slower moving—a combination of factors that is making them progressively more destructive. Throw in accelerating sea-level rise, due to more rapid melting of polar ice sheets, and the outlook for coastal cities looks especially grim. Storm records in the tropics continue to be smashed, the last decade seeing the most intense tropical cyclone on record, the most intense to make landfall, the most rapidly intensifying storm, the costliest tropical cyclone, and the costliest tropical cyclone season. It is unlikely that these records will stand for long.

Even though the global incidence of earthquakes and volcanic eruptions has remained unchanged as our once stable climate has begun to deteriorate, there are signs that the growing heat is beginning to exert its influence on the solid Earth too. In Alaska, where some of the glaciers have lost a vertical kilometre of ice in the last 100 years or so, the reduced load on active faults beneath is leading to a noticeable hike in earthquake activity. The melting of ice-cover on volcanoes in Iceland, the Andes, and the Cascade Range in western United States is also worrying, having the potential, as global heating continues to ramp up, to promote flank collapse and an increase in explosive eruptions.

According to the reinsurer, Aon, 'natural' hazards caused economic damage and disruption totalling a monumental US$323 *billion* in 2021, almost half the loss due to just three events—the collision of Hurricane Ida with the US state of Louisiana (US$75 billion), and flooding in Germany (US$46 billion) and China (US$30 billion). But these figures don't even begin to shed light on the scale of suffering and lost lives and livelihoods in countries where counting the monetary value of disaster means little. In 2021, Haiti was battered by the second major quake in eleven years, resulting in more than 2,200 deaths and leaving two-thirds of a million people desperately needing help. In the Philippines, the arrival of Super-typhoon Rai was endured by 13 million people, while in Indonesia, Cyclone Seroja brought severe flooding that destroyed tens of thousands of hectares of crops and more than 70,000 buildings.

Looking ahead, a combination of more people, increasing vulnerability and exposure, and progressive climate breakdown is set to make our world an even more perilous place. The population of the planet is slated to reach 10 billion around mid-century, after which the jury is out on whether or not it will continue to climb. Already, more people live in urban settings than in the countryside, and urban growth is set to continue. Most of Earth's new inhabitants will be born in Majority World countries, shoehorned into rapidly expanding metropolises in some of the planet's most dangerous locations.

We have plenty of means at our disposal to make communities more resilient, but these are often compromised by lack of political will, corruption, insufficient funds or expertise, and a focus on more immediate priorities, such as food security or housing. In a perfect world there should, in fact, be no reason for anyone to die as a consequence of a geophysical or climate hazard, at least in theory. The problem for us today is that the broader impacts on society and the economy of accelerating climate breakdown are going to make it even more difficult to build resilience. At the time we need it most, just coping with the growing bombardment of flood, storm, and wildfire, alongside earthquakes and volcanic eruptions, is going to leave progressively less time, effort, and funding to improve adaptation, and thereby increase resilience.

Whatever measures we take and wherever we live, we need to face the brutal fact that, due to accelerating climate breakdown, the world of our children and their children is certain to be a far more hostile one. Compared with someone born in 1960, a child born in a developed nation in 2020 will experience—on average—during the course of their lives, seven times more heatwaves, double the number of droughts, and three times as many harvest failures and floods. Inevitably, in Majority World countries, in the most vulnerable locations, these numbers will be considerably higher. And this already dismal picture is far from being a worst case. If we continue to do nothing to cut emissions, then the reality in the decades to come will be even more bleak. Challenging and enduring nature's shock troops, spearheaded by ever more extreme weather and rising seas, will, for many, become a daily ordeal rather than an occasional event. To say that adapting to the rapidly changing threat landscape of a planet whose climate is beginning to fail is going to be difficult is an understatement, but improving our knowledge and understanding of geophysical hazards, and how they are being modified and augmented by global heating, will go some way towards making the task a little more workable.

Bill McGuire

Brassington, Derbyshire
June 2023

List of illustrations

Geophysical and Climate Hazards

Chapter 1
Hazardous Earth

Danger: Nature at work

Our small planet has always been a dangerous place—more for some than others. In developed countries, the concentration of wealth means that the cost of geophysical catastrophe is almost inevitably totted up in dollars, euros, or pounds sterling. In Majority World nations, on the other hand, the concentration of people in cities in increasingly vulnerable locations means that the price of catastrophe is invariably measured in lives lost and livelihoods destroyed.

The growing and indiscriminate onslaught of extreme weather across the planet is, however, bringing increasing familiarity with lethal geophysical disasters to developed countries too. In this regard, the terrifying flash floods that took more than 240 lives across Belgium and Germany in 2021 were a real wakeup call, flagging the fact that if such devastation could reach the heart of a major western economy, then nowhere was any longer safe. Similarly, the unprecedented temperatures of summer 2022 brought to the UK and Europe a taste of the brutal heat that many in Majority World tropical nations now have to endure every year. They also contributed to the loss of more than 20,000, and possibly in excess of 50,000, lives across Europe, and approaching 3,000 over-65s in the UK. In Australia, New Zealand, and North

America, worsening heatwaves, wildfires, floods, and pumped-up storms are also taking an increasing toll on people's lives and on the economic and social fabric. The message is clear: no country, however advanced, is immune from the most severe climate-related hazards, and their impact is only going to grow in the decades ahead.

Meanwhile, irrespective of what's happening to the climate, the slow dance of the tectonic plates across the surface of our world continues as it always has, building strain on fault systems and releasing it in the form of often destructive earthquakes. Magma continues to rise from the uppermost mantle to feed reservoirs beneath and within active volcanoes, before accumulating pressures lead to either a violent explosion or the effusion of potentially destructive lava flows.

The first twelve years of the 21st century will long be remembered as the deadliest period of seismic activity in modern human history. During this time, all seven of this century's deadliest (so far) earthquakes have extinguished more than 600,000 lives across the Indian Ocean region and in Pakistan, China, Japan, Haiti, India, and Iran. Many of the lives have been lost to the devastating tsunamis triggered by the huge submarine earthquakes of 2004 (Indian Ocean) and 2011 (Japan), the like of which had never been experienced in living memory. In comparison—and notwithstanding lethal quakes in Nepal, Indonesia, and Haiti—the 10 years to 2022 have been relatively quiet, seismically speaking.

On the volcanic front, things have continued to rumble on with little in the way of major events, but plenty of spectacular shows of exploding and flowing lava, notably on the island of La Palma (Canary Islands) and at Fagradalsfjall on Iceland's Reykjanes Peninsula, both in 2021. Lives were lost during eruptions at Nyiragongo (Democratic Republic of Congo; DRC) in both 2002 and 2021, and at Merapi (Indonesia) in 2010. A collapse at the infamous Krakatau (Indonesia) volcano in 2018 generated a tsunami that took more than 400 lives, while other lethal eruptions occurred as far apart as New Zealand, Japan, Guatemala, the Philippines, and Eritrea.

From an impact perspective the biggest volcanic event of the 21st century to date was without question the relatively modest 2010 eruption of Iceland's Eyjafjallajökull volcano. Although no casualties resulted, an extensive ash cloud from the volcano caused aeronautical mayhem across the UK and Europe, leading to more than 100,000 flights being cancelled and costing the aviation industry an estimated US$1.7 billion.

The prize for the biggest volcanic surprise in recent years undoubtedly goes to the South Pacific submarine volcano, Hunga Tonga-Hunga Ha'apai, which blew itself apart in a mighty explosion in January 2022 (Figure 1). Imaged live by an orbiting satellite, the spectacular blast—the biggest this century—took few lives, but broadcast a moderate tsunami that reached the distant

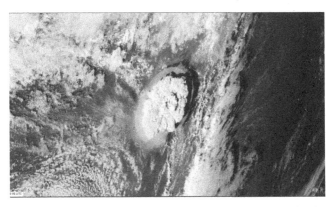

1. In January 2021, the Pacific volcano, Hunga Tonga-Hunga Ha'apai, exploded with the force of 2,500 Hiroshima atomic bombs, sending shock waves across the planet and generating sonic booms heard as far away as Alaska.

shores of Japan and the Americas, and spawned a supersonic shock-wave that circled the world.

The happenings of the first couple of decades of the 21st century just go to reinforce the fact that the Earth is an extraordinarily fragile place that is fraught with danger: a tiny rock hurtling through space at more than 100,000 kilometres an hour, wracked by violent movements of its crust and subject to dramatic climatic changes as its geophysical and orbital circumstances vary and its apex species does its worst. Barely 10,000 years after the end of the Ice Age, the planet is sweltering in some of the highest temperatures in at least 125,000 years as our once stable climate falters. At the same time, overpopulation, resource depletion, and environmental degradation are dramatically increasing the vulnerability of modern society to *all* geophysical and climate-related hazards.

For better and for worse, the Earth is the most dynamic planet in our solar system: a dynamism that has given us our protective magnetic field, our atmosphere, our oceans, and ultimately our lives. The very same geophysical features that make our world so life-giving and life-preserving also make it dangerous. The spectacular volcanoes that in the early history of our planet helped to generate the atmosphere and the oceans have, in the last three centuries, wiped out more than a quarter of a million people and injured countless others. At the same time, the rains that feed our rivers and provide us with the potable water that we need to survive have inundated enormous tracts of the planet with floods that in recent years have been truly biblical in scale. In any single year, floods impinge, on average, upon the lives of half a billion people and result in 25,000 deaths, and almost no country is any longer free of the threat.

I could go on in the same vein, describing how lives made enjoyable by a fresh fall of snow are swiftly ended when it avalanches, or how a fresh breeze that sets sailing dinghies

4

skimming across the wave tops can soon transform itself into a wailing banshee of terrible destruction—but you get the picture. Nature provides us with all our needs but we must be very wary of its rapidly changing moods, especially now that we have provoked it by dumping into the atmosphere more than 2.4 *trillion* tonnes of carbon dioxide in just a couple of centuries.

And on top of all this, constantly hanging over our heads like a Damoclean sword, is the menace of GGEs, or *global geophysical events*. These are natural phenomena writ large enough so that they have the potential to deal our civilization a serious, perhaps even fatal, blow. Fortunately, the GGE list is not long. At the top are large asteroid impacts and volcanic super-eruptions. With respective return periods of more than half a million years and 50,000 years, such events are rare indeed, so that the probability of one or other happening in any single year is vanishingly small. But they *will* happen—sometime. And there are others waiting in the wings too, less well known in popular culture perhaps, but with shorter return periods, and the potential to cause immense problems. These include ocean-wide mega-tsunamis projected by giant landslides at island volcanoes, and major earthquakes striking at one of the world's key trade or financial centres. Although the latter will not provide a geophysical shock at anything like a global scale, the economic consequences could be huge. Another societal and economic threat that has captured increased interest in recent years is that presented by a major solar storm, which has the potential to disable satellites and knock out power and communication networks on the ground.

What we mustn't forget, and admittedly this isn't likely given the evident impact of extreme weather in recent years, is that our world is already at the heart of a global geophysical event. Contemporary climate change due to anthropogenic global heating is the greatest threat our civilization and our race has ever had to face. It is now practically impossible for us to keep the global average temperature rise (since pre-industrial times) below

1.5°C—a figure widely regarded as marking the dangerous climate change guardrail. This means that we can no longer dodge perilous, all-pervasive, climate breakdown that will touch us all and insinuate itself into every aspect of our lives. The reality is that, if we don't act now to slash emissions, then dangerous will very quickly become cataclysmic. If we don't get on top of the climate emergency soon, when Yellowstone next erupts or the next large chunk of space debris slams into our world, we may not be here to care.

Earth: A potted biography

In order to understand geophysical hazards, and the ways in which many can be supercharged on an overheating world, it is essential to know a little about the Earth and how it functions. To begin, it is worth considering just how astonishingly old—at a little under 4.6 billion years—the Earth is. Old enough, in fact, to ensure that at some time in its history, anything that nature *can* conjure up, it already has. This includes everything from a worldwide magma ocean and a collision with another planetoid violent enough to gouge out the Moon, to a global snowball state and a climate balmy enough for crocodiles to bask at the poles.

And then there is life. Since the first single-celled organisms appeared on the scene billions of years ago, within sweltering chemical soups brooded over by a noxious atmosphere, life has struggled desperately to survive and evolve against a background of potentially lethal geophysical phenomena. Little has changed today, and the once natural perils that have assaulted our race in the past, and that constitute a growing future threat as we continue to destabilize our climate, have roots that extend back over four billion years to the creation of the solar system and the formation of the Earth from a disc of debris orbiting a primordial Sun.

During its early history, the Earth was pounded by chunks of rock left over from the formation of the eight planets, an extremely violent process that was mostly completed close to four billion years ago. Nonetheless, evidence for sporadic collisions between the Earth and *asteroids* and *comets*—respectively rocky and rock-ice bodies—can be found throughout our planet's geological record. Such comings together have been held responsible for a number of mass extinctions over the past half a billion years, including that which saw off the dinosaurs 66 million years ago. Furthermore, the threat of asteroid and comet impacts is still very much with us, and thousands of objects have been identified whose orbits around the Sun bring them a little too close to our world for comfort.

The primordial Earth bore considerably more resemblance to our worst vision of hell than today's stunning blue planet. The enormous heat generated by collisions, together with that produced by high concentrations of radioactive elements within the Earth, ensured that the entire surface was covered with a churning magma ocean, maybe 400 kilometres deep. Temperatures at this time were comparable with some of the cooler stars, perhaps approaching 5,000°C. Inevitably, where molten rock met the bitter cold of space, heat was lost rapidly, allowing the outermost levels of the magma ocean to solidify to a thin crust. Even before its 100 millionth birthday, however, the growing geological calm was shattered by the unimaginably violent coming together of the Earth and a Mars-sized body known as Theia; the resulting titanic collision scooping out a great chunk of our young world, forming around it a ring of debris that rapidly coalesced to make the Moon. By about 2.7 billion years ago more stable and long-lived crust managed to develop and to gradually thicken. Convection currents continued to stir in the hot and partially molten rock below. Consequently, the Earth's rigid outer layer was never a single, unbroken carapace, but instead broken up into separate rocky *plates* that moved independently on the backs of the sluggish convection currents.

Major changes were also occurring deep within the Earth's interior that, over time, developed a three-fold layered structure. At its heart, a composite core of iron and nickel—solid in the centre but liquid in its outer part—enclosed in a *mantle* of dense silicate minerals, partially molten in its uppermost part. Encasing this was the crust, made up of low-density, mainly silicate, minerals incorporated into rocks formed by volcanic action, sedimentation, and burial.

Above the Earth's surface, key changes were happening too. Our world's very earliest atmosphere was a noxious mixture of ammonia and methane, similar to those found today on Jupiter and Saturn. Volcanic outgassing on a prodigious scale saw nitrogen become dominant, before life began to introduce free oxygen to the atmosphere around 2.4 billion years ago. Despite some fluctuations in relative content, our planet's atmosphere has been dominated by nitrogen and oxygen ever since.

While only a relatively minor component of the atmosphere, carbon dioxide has always been present, and as a potent greenhouse gas, its variation over time has had important implications for the climate. Periods in our planet's history that saw relatively high carbon dioxide levels have also been characterized by warm, so-called *greenhouse*, conditions, which have prevailed across 85 per cent of Earth history. Low levels of the gas correlate with *icehouse* conditions, which have been dominant on five occasions since our world formed. We are currently living in the fifth of these—the Quaternary Ice Age—which began around 2.5 million years ago. In fact, we were already heading back into the next glacial period when the cooling trend was flipped by the huge quantities of carbon dioxide spewed out by human activities.

Over the last couple of billion years or so, things have quietened down considerably, and the geophysical and atmospheric processes that operate both within and at the surface of the planet have not changed a great deal. Ultimately, the hazards that constantly impinge upon our society result either from the Sun's incoming heat or from our planet's need to rid itself of the heat that is constantly generated in the interior by the decay of radioactive elements. The influence of the Sun's heat on the atmosphere and oceans drives our planet's weather machine that, as well as bringing balmy summer days and gentle winter snowfall, also unleashes devastating windstorms and biblical floods. In the Earth's interior, heat continues to be transported towards the surface by convection currents within the mantle. These are the engines that drive the great, rocky, plates across the surface of the planet, and underpin the concept of *plate tectonics*, which geophysicists use to provide a framework for how the Earth operates geologically.

The relative movements of the plates themselves, which comprise the crust and the uppermost rigid part of the mantle (together known as the *lithosphere*), are in turn directly related to the principal geological hazards—earthquakes and volcanoes—which are concentrated primarily along plate margins (see Figure 2). Here a number of interactions are possible. Two plates may scrape jerkily past one another, accumulating strain and releasing it periodically through destructive earthquakes. Probably the best known example of such a *conservative* plate margin is the quake-prone San Andreas Fault that separates western California from the rest of the United States.

Alternatively, two plates may collide head on. If they both carry continents built from low-density granite rock, as with the Indian Ocean and Eurasian plates, then the result of collision is the growth of a high mountain range—in this case the Himalayas— and at the same time the generation of major quakes such as that which wiped out 86,000 lives in Pakistan-administered Kashmir in October 2005. On the other hand, if an oceanic plate made of dense basalt hits a low-density continental plate then the former will plunge underneath its lighter companion, pushing back into the hot, convecting mantle. As one plate thrusts itself beneath the

9

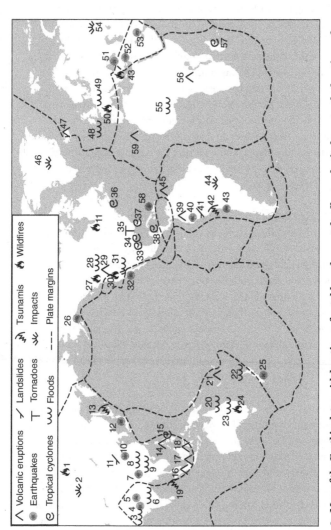

2. Map of the Earth's plates with locations of notable and recent deadly geophysical events: the locations of many geophysical disasters coincide with the plate margins.

1. Siberia Wildfires 2019 + 2021
2. Tunguska 1908
3. Pakistan Floods 2022
4. Kashmir 2005
5. Earthquake Nepal 2015
6. Bangladesh 1998
7. Sichuan 2008
8. Floods, China 2020
9. Yangtse 1998
10. Shensi 1556
11. Canada 2023
12. Kobe 1995
13. Tohuku 2011
14. Pinatubo 1991
15. Philippines 2013
16. Toba 74,000 BP
17. Krakatoa 1883
18. Tambora 1815
19. Sumatra 2004
20. Queensland 2011

21. Hunga-Tonga 2022
22. Floods Auckland 2023
23. Eastern Australia 2021
24. Wildfires Australia 2019 + 2020
25. Christchurch 2011
26. Alaska 1964
27. Canada Wildfire, 2021
28. Alberta 2013
29. Mt St Helens 1980
30. California Wildfires 2020 + 2021
31. California Floods 2022
32. Northridge 1994
33. Hurricane Texas 2017
34. New Orleans 2005
35. Missouri 2011
36. New York 2012
37. Miami 1992
38. Honduras + Nicaragua 1998
39. Nevado Del Ruiz 1985
40. Colombia 1995

41. Huascarán 1970
42. Chile 1960
43. Greece 2023
44. Brazil 1931
45. Mt Pelée 1902 and Soufriére Hills, Montserrat 1995
46. Greenland 1997
47. Eyjafjallajökull 2010
48. UK 2014
49. Germany + Belgium Floods 2021
50. European Wildfires 2021 + 2022
51. Izmit 1999
52. Earthquake Turkey 2023
53. Bam 2003
54. Chelyabinsk 2013
55. Nigeria Floods 2022
56. Nyiragongo 2021
57. Cyclone Madagascar 2019 + 2022
58. Earthquakes Haiti 2010 + 2021
59. La Palma 2021

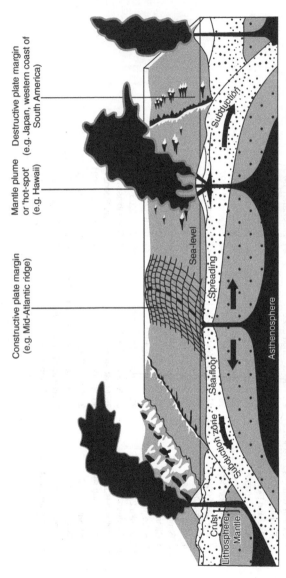

Constructive plate margin (e.g. Mid-Atlantic ridge)

Mantle plume or 'hot-spot' (e.g. Hawaii)

Destructive plate margin (e.g. Japan, western coast of South America)

Sea-level

Sea-floor

Spreading

Subduction zone

Subduction

Subduction

Asthenosphere

Crust
Lithosphere
Mantle

3. The lithosphere, the Earth's outer rigid shell, is created at mid-ocean ridges and destroyed in subduction zones, where most of the Earth's deadliest earthquakes and volcanic eruptions occur.

other (a process known as *subduction*; see Figure 3) so the world's greatest earthquakes are generated, including the tsunami-generating quakes of Sumatra (Indonesia) in 2004 and northern Japan in 2011. Subduction is happening all around the Pacific Rim, ensuring high levels of seismic activity in Alaska, Japan, Taiwan, the Philippines, Chile, and elsewhere in the circum-Pacific region. This type of *destructive* plate margin—so called because one of the two colliding plates is consumed—also hosts large numbers of active volcanoes.

Although the mechanics of magma formation in such regions is sometimes complex, it is ultimately a result of the subduction process and owes much to the partial melting of the subducting plate as it is pushed down into ever hotter levels in the mantle. Fresh magma formed in this way rises as a result of its low density relative to the surrounding rocks, and blasts its way through the surface at volcanoes that are typically explosive and, therefore, especially hazardous. Strings of literally hundreds of active and dormant volcanoes—which host nearly all the world's large, lethal, eruptions—circle the Pacific, making up the legendary *Ring of Fire*, while others squat above subduction zones in the Caribbean and Indonesia.

To compensate for the consumption of some plate material, new rock must be created to take its place. This happens at so-called *constructive* plate margins, where two plates are moving away from each other, and along which fresh magma rises from the mantle to fill the space. This occurs mainly beneath the oceans along a 40,000 kilometres long network of linear topographic highs known as the *Mid-Ocean Ridge system*, where newly created lithosphere exactly balances that which is lost back into the mantle at destructive margins (see Figure 3). A major part of the Mid-Ocean Ridge system runs down the middle of the Atlantic Ocean, bisecting Iceland, and separating the Eurasian and African plates in the east from the North American and South American plates in the west. Here too there are both volcanoes and earthquakes, but the former generally involve relatively mild eruptions and the latter are small.

Driven by the mantle convection currents beneath, the plates waltz endlessly across the surface of the Earth, at about the same rate as fingernails grow, constantly modifying the appearance of our planet and ensuring that, given time, everywhere gets its fair share of volcanic eruptions and earthquakes.

Perilous planet

At any single point, and at any one time, the Earth and its enclosing atmospheric envelope give the impression of being mundanely immutable and benign. This is, however, an entirely misleading notion, with something like 3,500 detected earthquakes agitating the planet every day and a volcano erupting every week or so. Each year, the tropics are battered by up to 40 hurricanes, typhoons, and cyclones, while floods and landslides occur everywhere in numbers too great to keep track of.

Of the so-called geological hazards—earthquakes, volcanic eruptions, and landslides—there is no question that earthquakes are by far the most devastating. Every year about 150 quakes reach or exceed magnitude 6, making them large enough to cause significant damage and loss of life, particularly when they strike poorly constructed and ill-prepared population centres in Majority World countries. As previously mentioned, most big earthquakes are confined to distinct zones that coincide with the margins of plates. This is where two of the most notorious quakes of modern times wrought their destruction: the 2004 Sumatra (Indonesia) earthquake and tsunami taking 230,000 lives across the region, and the 2011 Tohoku (Japan) earthquake and tsunami, qualifying—at US$235 billion—as the costliest disaster in history.

Most of the colossal death tolls arising from major earthquakes are a consequence of the collapse of buildings that are simply not up to the job of protecting those inside, but earthquakes kill in other ways too. Due to the Sumatra and Tohoku events, we are all now far more familiar with the tsunami than we were even a quarter of a century ago. These extreme waves are generated when a submarine quake instantaneously jerks upwards, sometimes by several metres, a large area of the seabed, causing the displaced water above to hurtle outwards as a series of ripples. In deep water, these ripples may be only a few tens of centimetres high, but when they enter shallower water they build in height, maybe to 30 metres or more, before crashing into coastal zones with extreme force. The powerful ground shaking that accompanies a large earthquake is also effective in triggering the collapse of unstable slopes, in the worst cases at a colossal scale. During the 1920 Haiyuan earthquake (China), up to 50,000 landslides are estimated to have killed more than 32,000 people out of a total death toll of around 234,000.

Estimates of the number of active volcanoes vary, but there are at least 1,500 and possibly more than 3,000. Every year around 70 volcanoes erupt, some of which—like Kilauea on Hawaii or Stromboli in Italy—are almost constantly active. Others may have been quiet for centuries or, in some cases millennia, and eruptions at these tend to be especially dangerous. The most violent volcanoes occur at destructive plate margins, where one plate is consuming another. Their outbursts rarely produce quiet flows of lava and are more likely to launch enormous columns of ash and debris 20 kilometres or more into the atmosphere. Carried by the wind over huge areas, volcanic ash can be extremely disruptive, making travel difficult, damaging buildings, electronics, and crops, poisoning livestock, contaminating water supplies, and causing health problems. Thick ash deposits can also provide a source for mudflows, triggered by heavy rain. Almost a decade after the 1991 eruption of Pinatubo (Philippines), mud pouring off the volcano was still clogging rivers, inundating towns and agricultural land, and damaging fisheries and coral reefs. Somewhat surprisingly, mudflows also constitute one of the biggest killers at active volcanoes. In 1985 a small eruption through the ice and snow fields of Columbia's Nevado del Ruiz volcano unleashed a torrent of mud out of all proportion to the size of the eruption, and took 23,000 lives.

Even scarier and more destructive than volcanic mudflows are *pyroclastic flows* or 'glowing avalanches'. These hurricane-force blasts of incandescent gas, hot ash, and blocks and boulders sometimes as large as houses, have the power to obliterate everything in their paths. And the threat from volcanoes doesn't end there: chunks of rock collapsing into the sea from their flanks can trigger huge tsunamis, while noxious fumes can and have killed thousands along with their livestock. Volcanic gases from major explosive eruptions, carried into the stratosphere and from there around the planet, have modified the climate and led to miserable weather, crop failures, and health problems half a world away. On the grandest scale, volcanic super-eruptions have the potential to affect us all, plunging the planet into a frigid *volcanic winter* and devastating harvests worldwide.

Of all geological hazards, landslides are perhaps the most underestimated, probably because they are often triggered by some other hazard, such as an earthquake or deluge, and the resulting damage and loss of life is therefore subsumed within the tally of the primary event. Nevertheless, landslides can be highly destructive, both in isolation and in numbers. In 1970, a moderate quake caused the entire peak of the Nevados Huascaran mountain in the Peruvian Andes to fall on the towns below, snuffing out the lives of 18,000 people in just four minutes, and erasing all signs of their existence from the face of the Earth. Heavy rainfall too can be particularly effective at triggering landslides and when, in 1998, Hurricane Mitch dumped more than a metre of rain on Central America in 36 hours, it mobilized more than a million landslides in Honduras alone, blocking roads, burying farmland, and destroying communities.

While earthquakes and volcanic eruptions—and to some extent landslides—are linked to how our planet functions geologically, other geophysical hazards are more dependent upon processes that operate in the Earth's atmosphere. Rather than by heat from the interior, our planet's weather is driven by energy from the Sun. But this is now augmented by the thickening blanket of greenhouse gases in the atmosphere, reflecting more than two centuries of polluting human activities. Our nearest star is the ultimate instigator—aided by the Earth's rotation and the constant exchange of energy and water with the oceans—of the tropical cyclones and floods that exact an enormous toll on life and property, particularly in Majority World countries. The growing intensity and frequency of such events, however, is a direct consequence of global heating and ongoing climate breakdown.

Still other lethal natural phenomena have a composite origin and are less easy to pigeonhole. As well as being triggered by earthquakes, tsunamis, for example, can be formed by landslides into the ocean and by eruptions of coastal and island volcanoes. Similarly, many landslides result from collusion between geology and meteorology, with torrential rainfall destabilizing already weak slopes.

In terms of the number of people affected, floods undoubtedly constitute the greatest of all geophysical hazards. Since the start of the century, the fraction of the global population at risk of flooding has climbed by almost a quarter, largely driven by a failing climate and more extreme weather. Between 2000 and 2018, up to 300 million people were impacted by floods, a figure that is only going one way, given a future of rising sea levels and more extreme precipitation. River floods are respecters of neither wealth nor status, and both developed and Majority World countries have been severely afflicted in recent years, across every continent. Wherever rain is unusually torrential or persistent, it will not be long before river catchments fail to contain surface

4. Global heating made the torrential rainfall that caused the 2022 Pakistan floods 50 per cent more intense.

run-off and start to expand across their flood plains and beyond. As flood plains all over the world become more crowded, and nonsensically covered in concrete and tarmac, the loss of life and damage to property caused by swollen rivers has increased dramatically. The astonishing scale of the 2022 Pakistan floods, which inundated fully 12 per cent of the country, bears testament to just how devastating these events can be on an overheating planet (see Figure 4).

Partly through their effectiveness at spawning floods, but also through the enormous wind speeds achieved, storms constitute one of the most destructive of all geophysical hazards. Furthermore, because they are particularly common in some of the world's most affluent regions, they are responsible for some of the most costly geophysical disasters of all time. Every year, the Caribbean, the Gulf of Mexico, and the southern states of the USA, along with parts of South and South-East Asia are struck by tropical cyclones, while the UK and continental Europe suffer from severe and damaging winter storms. In the tropics, storms get their energy from the warm surface waters. As global heating drives up sea-surface temperatures, it should come as no surprise that the most powerful hurricanes and typhoons are becoming increasingly prevalent, growing in their potential for destruction if and when they make landfall. Outside the tropics, at mid-latitudes, the number of the most powerful so-called *extra-tropical* cyclones is projected to increase too, even as the total number of such storms may fall. Those of us who remember the devastating Great Storm of October 1987 in the UK will appreciate that this is not good news. The jury is still out on how global heating might affect those rotating maelstroms of solid wind known as tornadoes, although some changes have already been observed in the US where tornado outbreaks have, in recent decades, become larger and more frequent.

The final, and arguably the greatest, threat to life and limb comes not from within the Earth but from beyond the atmosphere. Although the near constant bombardment of our planet by large chunks of space debris ended billions of years ago, the menace from comets and asteroids remains real enough for NASA to take potshots at one to test diversionary tactics. Recent estimates suggest that around 1,000 asteroids with diameters of 1 kilometre or more have orbits around the Sun that approach or cross the Earth's, making collision possible at some point in the future: This population includes many objects 2 kilometres across and larger. An object of this size striking our planet would trigger a *cosmic winter*, due to dust lifted into the stratosphere blocking out solar radiation, with dire consequences for global society and its economy.

Hazards on a hotter world

We only need to look around us to realize that a hotter world with a destabilized climate is also a more perilous one, but this is supported by more than just a subjective personal perspective. For decades, climate change deniers, could—and did—scupper any suggestion that a particular drought, heatwave, flood, or storm was a consequence of global heating, by pronouncing that such extreme weather was nothing new and had always been with us. Now, however, this particular avenue of obfuscation has been closed off. Today, *attribution analysis* is used to determine just how likely a particular extreme weather event would have been in the absence of global heating, so it is possible to begin to build an objective model of how our once stable climate is unravelling. The methodology takes account of the frequency of occurrence and intensity of a particular event over time, and compares these parameters with those determined through modelling the same event in a world with elevated atmospheric greenhouse gas levels causing global heating. The blistering heatwave that smashed temperature records across the UK in summer 2022, for example, is shown in this way to have been made at least 10 times more likely by global heating due to human activities. Similarly, the intensity of the rainfall that spawned the calamitous Pakistan floods in the same year was increased by up to 50 per cent by global heating (Figure 4).

In fact, a wide-ranging and forensic analysis of more than 500 extreme weather events undertaken by the well-respected scientist- and journalist-run UK website, Carbon Brief, for *The Guardian* newspaper, revealed that the fingerprints of global heating are all over them. At the top of the pile, a dozen events—including intense heatwaves in North America, Japan, and Europe—were shown to have been impossible in the absence of global heating. Almost three-quarters of all events were demonstrated to have been made either more likely or more severe due to climate breakdown caused by human activities. Probably the most astonishing outcome of the study was the revelation that one in three heat-related summer deaths in the last 30 years—a figure in the millions—were a direct result of global heating. There is, then, no denying the brutal truth that, in continuing to pump out vast quantities of carbon dioxide, we are making a rod for our own backs and for the backs of generations to come. We are whistling up a whirlwind of extreme weather that millions, if not billions, will reap.

Geophysical hazards and us

Looking back in time, the human cost of geophysical hazards is colossal. The reinsurance company, Munich Re., estimates (rather conservatively I suspect) that up to 15 million people were killed by 'natural' hazards in the last millennium, and over 3.5 million in the 20th century alone. The most lethal year in recent decades was 2004, during which the Indian Ocean tsunami, together with earthquakes in Morocco and Japan, record storms in the US and Japan, and flooding across Asia contributed to the loss of one-third of a million lives. In terms of economic cost, the pinnacle year remains 2011, which, as well as the Japan tsunami, saw earthquakes in Turkey and New Zealand; severe floods in Thailand, Australia, China, and Brazil; a massive tornado outbreak in the United States; and storms in Korea, Japan, the Philippines, the Caribbean, and western Europe. These added up to damage—according to US reinsurer Aon—totalling US$615 billion. This seemingly exceptional annual toll will undoubtedly, however, be far exceeded as the impacts of climate-related catastrophes continue to ramp up. These contributed substantially to the combined losses for the decade to 2020 of US$3 *trillion*—a trillion more than the previous decade. In 2021, fully 96 per cent of losses were weather related, flagging their overwhelming dominance on the disaster front.

Hazardous geophysical events impacted on more than 2.5 billion people in the first decade of the 21st century, while the Red Cross estimates that in the decade to 2020, an astonishing 1.7 billion people were affected by climate-related disasters that took 410,000 lives. Unhappily, there is no sign that hazard impacts on society are diminishing as a consequence of improvements in forecasting and mitigation or attempts at building resilience, and figures suggest that we are losing the battle against the impacts of accelerating global heating and nature's dark side. While we now know far more about geophysical hazards, the mechanisms that drive them, and their sometimes terrible consequences, any benefits accruing from this knowledge have been more than negated by the impact of rising temperatures and extreme weather, and the increased vulnerability of large sections of the Earth's population.

The latter has arisen primarily as a consequence of the precipitous rise in the number of people on the planet, which doubled between 1960 and 2000 and has recently topped the eight billion mark. The bulk of this rise has occurred in poor Majority World countries, many of which are susceptible to a whole spectrum of geophysical hazards. Furthermore, the struggle for living space has ensured that marginal land, such as steep hillsides, flood plains, and coastal zones, has become increasingly utilized for farming and habitation. Such terrains are clearly high risk and can expect to succumb on a more frequent basis to, respectively, landslides, flooding, storm surges, and tsunamis.

As mentioned earlier, another major factor in raising vulnerability in recent years has been the move towards urbanization in the most hazard-prone regions of the planet. In 2007, for the first time ever, the number of people living in urban environments overtook the number residing in the countryside, and in 2021, this figure reached 57 per cent. Most of the new city-dwellers are crammed into poorly sited and badly constructed *mega-cities*, typically defined as those with populations in excess of 10 million. Forty years ago, New York and London topped the league table of world cities with populations, respectively, of 12 and 8.7 million. Cities today, however, are on a different scale entirely (Table 1). Tokyo is top of the pile, with a metropolitan population in excess of 37 million, with Delhi second on close to 30 million. Majority World cities, such as São Paolo (Brazil), Dhaka (Bangladesh), Cairo (Egypt), Mexico City, and Mumbai (India) are all now gigantic sprawling agglomerations of humanity with populations

Table 1. The world's biggest urban areas (millions of inhabitants)

City	Country	UN estimate 2018	City proper	Urban area
Tokyo	Japan	37.5	13.5	37.7
Delhi	India	28.5	16.7	32.2
Shanghai	China	25.6	24.9	24.0
São Paolo	Brazil	21.6	12.3	23.1
Mexico City	Mexico	21.6	9.2	21.8
Cairo	Egypt	20.1	9.5	20.3
Mumbai	India	20.0	12.5	25
Beijing	China	19.6	21.9	18.5
Dhaka	Bangladesh	19.6	8.9	18.6
Osaka	Japan	19.3	2.7	15.1

Sources: United Nations and Demographia World Urban Areas, 18th edn, July 2022.

exceeding 20 million, and extremely vulnerable to geophysical hazards. These great conurbations will bear the brunt of nature's savagery—in the form of earthquake, tsunami, or volcanic blast—pumped up further by anthropogenic global heating, bringing blistering heatwaves, increasingly powerful storms, and more extreme flooding.

A staggering 96 per cent of all deaths arising from geophysical hazards and environmental degradation occur in Majority World nations, and there is no prospect of this figure falling. Indeed, as climate breakdown bites ever harder, this bleak picture is certain to deteriorate even further. In particular, summer heat is going to become an increasingly serious problem, compounded by the so-called 'urban heat island' (UHI) effect. The acres of concrete and tarmac, industrial activity, pollution, and the sheer mass of people and vehicles can make cities several degrees Celsius hotter than the surrounding countryside. Factor in rising levels of humidity and the almost complete absence of air conditioning, and this is a recipe for major loss of life during extended intense heatwaves. Despite efforts to build more resilient communities, with so many people shoehorned into ramshackle and dangerously exposed cities, many in coastal locations at risk from earthquakes, tsunamis, storms, and rising sea levels, it can only be a matter of time before we see the first of a series of true *mega*-disasters, with death tolls exceeding one million.

Chapter 2
Earthquakes and tsunamis

The shaking Earth

Earthquakes are our planet's way of releasing strain accumulated as a consequence of the relative movements of the tectonic plates. But the by-blow is death and destruction, sometimes on a massive scale. During the last millennium, earthquakes were responsible for the loss of at least eight million lives. Between 1998 and 2017 alone, around a quarter of a million deaths were caused by quakes, with an estimated 150 million affected in some way. Averaged over time, about 20,000 people are killed by earthquakes every year, making them the most lethal of all geophysical hazards. Terrifying as these figures sound, the rapid growth of mega-cities in regions of high seismic risk mean that death tolls are likely to climb, rather than fall, in the decades ahead.

The surface of our planet is never still and small earthquakes can, and do, happen almost anywhere at any time. Every year more than half a million are sufficiently energetic to be detected by seismometers and maybe a fifth of these shake the ground enough to be felt by humans. On average, however, little more than 100 or so are violent enough to cause loss of life and significant damage if they happen in the wrong place. Of these, only about a dozen are termed *major* quakes—between magnitude 7 and 7.9—while so-called *great* earthquakes—defined as magnitude 8 or above—happen just once a year at most. Every now and again we get a magnitude 9+ event, but these are so infrequent that just five have been recorded since the beginning of the 20th century. While very rare these prodigious quakes are important because they release most of the world's pent-up seismic energy.

The amount of energy released in an earthquake is closely related to the *magnitude* of the quake, which is a measure of its size or strength. In addition to the magnitude scale developed by Charles Richter and collaborator Beno Gutenberg in 1935—and known as the *Gutenberg-Richter Scale*—there are now a plethora of others. Different scales may be used according to the circumstances surrounding a particular quake, but the most authoritative in use today is the *Moment Magnitude Scale*. This has a number of advantages, including the fact that it is most closely linked to the energy released by an earthquake, and does not 'saturate' towards the top of the scale (as Gutenberg-Richter does), making it more effective at differentiating the sizes of great earthquakes. As such, the Moment Magnitude Scale—denoted by Mw—is the standard scale used by the United States Geological Survey (USGS) for reporting earthquakes above magnitude 4.

A key point about all magnitude scales is that they are logarithmic rather than linear, so that every step up in magnitude involves a 10 times increase in amplitude (a measure of the ground motion) as determined from a seismograph. Each step up also involves a 32 times increase in energy release. A magnitude 7 quake, for example, would release 1,000 times more energy than a magnitude 5. To provide some idea of what this means in terms of the energy available for damage and destruction, a magnitude 1 quake would produce the same amount of energy as blowing up 170 grams of TNT, while a magnitude 8 quake would release energy equivalent to detonating six *million* tonnes of the explosive.

All magnitude scales are open-ended and range up to just over 9, the biggest quake ever recorded—the Valdivia earthquake that struck off the coast of Chile in 1960, scoring an Mw of 9.5. This extraordinary event released a quarter of all the world's seismic energy for the period 1900–2005, and together with the 1964 Alaska quake (Mw = 9.4) and 2004 Sumatra quake (Mw = 9.3), almost half of all the seismic energy over this period. It is worth making an important observation here, which is that a magnitude 10 earthquake, as featured in a number of invariably awful disaster B-movies, is not possible. This is because the amount of energy released by an earthquake is strongly related to the length of the fault that has ruptured to produce it. To generate an earthquake of Mw = 10, a fault 14,000 kilometres long would need to rupture in one go. No fault of this length exists anywhere on the planet. Even if we added up the energy released by *all* the earthquakes that happened over the last 110 years, this would still only add up to an Mw of 9.95.

Magnitude scales are very useful for determining earthquake parameters and helping to understand mechanisms, but they only provide a guide to the destructive capacity of an earthquake. Far better for this is a measure of the amount or intensity of shaking. In simple terms, the more vigorously the ground shakes, the more damage and destruction can be expected, but, for a given magnitude quake, the intensity of shaking will depend on a number of factors. Foremost amongst these is the distance from the earthquake epicentre—as, all other things being equal, shaking diminishes with distance. Other factors include the alignment of the fault rupture (e.g. north–south, east–west, etc.) relative to a particular location, the depth of the earthquake, and the nature of the underlying geology, which can be crucial. Soft sediment, for example, can undergo so-called *liquefaction* due to ground shaking, causing it to behave as a liquid and leading to the foundering of buildings and other structures. Unconsolidated sediment can also magnify shaking several times and extend its duration, greatly increasing the damage potential of a quake. Local variations, known as *site effects*, in what lies beneath the surface can result in a row of buildings on one street being flattened, while those on a neighbouring street remain untouched.

A number of different scales are available for measuring earthquake intensity, but the most commonly utilized is the *Modified Mercalli Scale* (MMS) (Table 2), which was developed in 1902 by the Italian volcanologist and Catholic priest Giuseppe Mercalli, and modified in the 1930s. Unlike magnitude scales, intensity scales are subjective measures of the effects and damage arising from an earthquake, and are based on observations such as whether shaking woke people up, moved furniture, brought down chimneys, or caused masonry buildings to collapse. To avoid confusion with magnitude measurements, MMS intensities are denoted by Roman numerals, ranging from I—shaking not felt—up to X—extreme shaking, resulting in most masonry buildings being destroyed or severely damaged. Two additional categories, XI and XII, have been proposed for the top end of the scale, but these are usually merged into a single X+ category, marking an exceptional level of destruction.

One added benefit of the MMS is that it can be used to paint a picture of ground shaking in historic earthquakes that occurred before the arrival of seismometers, but for which there are reasonable descriptions of the effects of the event. There are drawbacks too, a key one being that it only works for inhabited areas. Without any people or buildings to shed light on the impact of a quake it is impossible to determine its intensity using the scale. Furthermore, the MMS says nothing about the strength of the earthquake at its source. For example the damage experienced by a particular settlement would be the same whether it was caused by a medium-sized quake nearby or a bigger quake further away.

Because magnitude allocation is based on interpretation of the seismic record, and intensity determination depends on observation, which can be subjective, direct comparison of the two

Table 2. The Modified Mercalli Scale

Intensity	Shaking	Effects/damage
I	Not felt	Not felt except by very few under especially favourable conditions.
II	Weak	Felt only by a few persons at rest, especially on the upper floors of buildings.
III	Weak	Felt quite noticeably by persons indoors, especially on the upper floors of buildings. Stationary motor cars may rock slightly. Vibrations similar to a passing truck.
IV	Light	Felt indoors by many, outdoors by few during the day. At night some awakened. Dishes, windows, doors disturbed; walls make cracking sounds. Sensation like heavy truck striking a building. Stationary motor cars rocked noticeably.
V	Moderate	Felt by nearly everyone, many awakened. Some dishes, windows broken. Unstable objects overturned. Pendulum clocks may stop.
VI	Strong	Felt by all, many frightened. Some heavy furniture moved; a few instances of fallen plaster. Damage slight.
VII	Very strong	Damage negligible in buildings of good design and construction; slight to moderate in well-built ordinary structures; considerable damage in poorly built or badly designed structures; some chimneys broken.
VIII	Severe	Damage slight in specially designed structures; considerable damage in ordinary substantial buildings with partial collapse. Damage great in poorly built structures. Fall of chimneys, factory stacks, columns, monuments, walls. Heavy furniture overturned.
IX	Violent	Damage considerable in specially designed structures; well-designed frame structures thrown out of plumb. Damage great in substantial buildings with partial collapse. Buildings shifted off foundations.

Earthquakes and tsunamis

(continued)

Table 2. Continued

Intensity	Shaking	Effects/damage
X	Extreme	Most masonry and frame structures destroyed; some well-built wooden structures destroyed. Rails bent.

Notes: The Modified Mercalli Scale measures the intensity of seismic shaking. Sometimes two extra levels (XI and XII) are used to deal with exceptional levels of destruction.

is a bit like comparing apples and pears—not possible. Although build quality is a complicating factor, some broad correlations can, however, be made so that it becomes possible to derive ball-park magnitude estimates for earthquakes that happened prior to the advent of a global seismic network. Potentially destructive earthquakes of magnitude 6–6.9, for example, would be expected to result in damage that would register at VIII or above on the MMS. Major quakes—those recording a magnitude of 7–7.9—would, meanwhile, clock in at X or X+.

Quakes and plates

While the ground *can* shake almost anywhere, the biggest, and therefore most destructive and lethal, earthquakes are confined to destructive, collisional, or conservative plate boundaries, mainly in a belt that coincides with the Pacific Rim and in another that extends east from southern Europe, through Turkey, Iran, central and southern Asia, and on into China. In the first couple of decades of the 21st century, nine out of the 10 deadliest earthquakes occurred within these two seismic belts, the exception being the 2010 Haiti quake. Earthquakes do happen at constructional boundaries, where two plates are moving apart, but these tend to be small and are—barring those that periodically strike Iceland—typically located beneath the ocean floor and remote from population centres. So-called *intraplate* earthquakes, which happen far from plate margins, are rare, but can be damaging, or even destructive. Probably the best known intraplate earthquake, which was in fact a series of quakes, shook the US state of Missouri in 1811–12 close to the small settlement of New Madrid. The series included three events larger than Mw = 7 and caused damage across an area of 600,000 square kilometres. Concerns remain in the area about a repeat of the event, about which more later.

The c.10,500 kilometre long seismic belt that stretches from southern Europe to China marks the position of a zone along which the African, Arabian, and Indian plates to the south, along with a number of smaller platelets, crashed, over the course of the last 50 million years or so, into the giant Eurasian plate to the north. As well as pushing up mountain ranges, including the Pyrenees, Alps, and Himalayas, the northward movement, which still continues today, also resulted in the formation of numerous fault systems. These continue to accumulate strain, periodically releasing it by generating earthquakes that may exceed Mw = 8. Recent examples of devastating earthquakes occurring in this belt include the 7.7 magnitude Gujarat quake (2001), 7.6 magnitude Kashmir (Pakistan) event (2005), the 7.9 Sichuan earthquake (2008), the 7.8 Nepal quake (2015), and others in Iran, Albania, Afghanistan, Azerbaijan, Greece, Iran, and elsewhere. Going back in time, a magnitude 8.6 earthquake devastated the Indian state of Assam in 1950, while the 7.6 quake that struck the Chinese region of Tangshan in 1976 took up to 300,000 lives.

Earthquakes on conservative plate boundaries, where two plates are sliding laterally past one another, are caused by the fault plane progressively locking, accumulating strain over decades or centuries, and then releasing it spontaneously. Probably the two best known examples of such a boundary are California's San Andreas Fault; and the North Anatolian Fault, which extends right across northern Turkey, paralleling the southern coast of the Black Sea. Vying with Florida for the title of the *Sunshine State*, California can justifiably be described as the 'earthquake state', hosting as it does more than 7,000 quakes every year. As I write this (in mid-June 2023), the state has experienced 23 earthquakes exceeding magnitude 1.5 in just the last 24 hours.

The reason California is shaken so often is the presence of the 1,200 kilometres long San Andreas Fault and its many subsidiaries. The fault marks the boundary between the northward-moving Pacific Plate to the west and the southward-moving North American Plate to the east. The average annual slip on the fault is around 35 millimetres. Since its formation, believed to be around 15–20 million years ago, accumulated slip has shifted the Pacific Plate north by more than 500 kilometres. At this rate—millions of years hence—fault movement will carry Los Angeles, located to the west of the main fault, northwards past San Francisco. The San Andreas has hosted numerous damaging earthquakes and a few deadly ones. Foremost amongst these is the magnitude 7.9 San Francisco quake of 1906. Intensity X+ shaking resulted in severe damage, which was made worse by a firestorm that consumed many of the mostly wooden buildings (Figure 5). More than 80 per cent of the city was destroyed and in excess of 3,000 lives lost. Nearer to the present time, serious earthquakes in 1989 (Loma Prieta) and 1994 (Northridge) caused significant damage although the casualty figures were low.

Longer—at 1,500 kilometres—than the San Andreas, Turkey's North Anatolian Fault, which separates the Eurasian Plate to the north from the Anatolian Plate in the south, has proved to be far more lethal than its Californian counterpart. Since the 1930s the fault has been 'unzipping' from east to west in a series of earthquakes—seven of them of magnitude 7 or above—that together have taken more than 50,000 lives. The deadliest of these were the 1939 magnitude 7.8 Erzincan event, in the east of the country, which resulted in the loss of more than 32,000 lives, and the 1999 magnitude 7.6 quake, immediately west of Istanbul, which caused more than 17,000 deaths. Both of these disasters were, however, dwarfed by the devastating earthquakes that

32

5. **The burned-out ruins of San Francisco following the 1906 earthquake and firestorms.**

struck southern Turkey and northern Syria in February 2023. Twin quakes of magnitude 7.8 and 7.7, caused by the rupturing of the East Anatolian Fault, which marks the contact of the Anatolian Plate to the north and the Arabian Plate to the south, resulted in serious damage across an area of more than one-third of a million square kilometres, and took around 60,000 lives.

More than 90 per cent of the world's destructive plate boundaries are located around the margins of the Pacific Ocean basin. Here, subduction zones in the north and east mark where the Pacific and other smaller plates slide beneath North and South America and, in the south and west, beneath the Eurasian, Philippine, and Australian plates. A salient extends westwards into the eastern Indian Ocean, where the Indo-Australian Plate plunges down beneath the islands of Indonesia. Close to 90 per cent of all earthquakes happen around the Pacific Rim, releasing more than three-quarters of our world's seismic energy. It is here too that the greatest earthquakes occur, including 9 out of 10 of the biggest quakes of the last 100 years.

The colossal earthquakes that shake the Pacific Rim on an all too frequent basis occur beneath the sea on so-called *mega-thrusts*. These are major, low-angle faults, often more than 1,000 kilometres in length, which mark the contact between the overriding plate and the plate sliding beneath. Typically, the former snags on the subducting plate, with its leading edge pulled down as a result and slowly accumulating strain, often over periods as long as centuries. When a critical threshold is reached, the locked plates free themselves, releasing the accumulated strain in a devastating burst of seismic energy. At the same time, the leading edge of the deforming plate pings upwards to resume its former position, imparting an impulse to the ocean above. This generates a bulge of water that subsides to a series of ripples that speed out in opposing directions perpendicular to the length of the fault. When and if they approach land, they grow rapidly in height to become the tsunamis that can magnify enormously the lethal and destructive capacities of the quakes that spawned them.

Both the Sumatra (2004) and Tohoku (2011) earthquakes and tsunamis were mega-thrust events, alongside the world's greatest ever recorded quake at Valdivia (Chile) in 1960, and the second largest off Alaska in 1964. Both of these also generated lethal ocean-wide tsunamis, taking at least 1,000 lives in 1960, including more than 60 in Hawaii, and a further 120 in 1964. Going a little further back in time, another great quake worth a mention is that caused in 1700 by the rupture of the Cascadia Subduction Zone, which stretches for 1,000 kilometres off the west coast of North America, from Canada's Vancouver Island to northern California. Estimated at around magnitude 9, this event also generated a major tsunami, which is recorded in indigenous American legends and in written reports in Japan. The threat of a future great earthquake in the Cascadia Subduction Zone is once again increasing, and I will say a bit more about this shortly.

Earthquake prediction—Holy grail or poisoned chalice

The idea persists in some circles that all would be well if we could predict earthquakes accurately, so that warnings could be issued meaning that no-one would need to die. But would this, in reality, be a good thing? Conversations about earthquake prediction are often confused by the fact that the terms earthquake forecasting and earthquake prediction are often used interchangeably, which is both wrong and misleading. This is because, while we can now forecast earthquakes, we are no closer to predicting them than we were half a century or more back.

Because the accumulation and release of strain on a particular fault is reasonably consistent over time, earthquakes tend to happen with a rough periodicity. One fault, for example, may rupture—on average—about once every century, while another might only generate a quake every 1,000 years or so. As such, earthquakes on a specific fault have average *return periods* that can be used as guides to when the next one might be due. If, for example, a fault ruptures, on average, every 80 years, and there has not been one for 90 years, then it would be reasonable to say that one could be expected soon. This, however, is a forecast—not a prediction. Broadly speaking, the longer the record, the more accurate a forecast based on it is likely to be. For the Los Angeles (LA) area of California such a record is available from the dating of earthquake-related deformation over the past 1,500 years or so. This reveals that since 565 AD there have been eight major earthquakes, spaced at intervals ranging from 55 to 275 years, and with an average return time of 160 years. The last time the Earth moved in a big way was in 1857, so LA has a reasonable chance of being faced with the next 'big one' sometime in the near to medium term. Then again, it might not happen until 2132. As time goes on, percentage probabilities of the occurrence of the next quake can also be made. In the case of Los Angeles, for example, the USGS calculates that the probability of a magnitude 6.7 quake in the next 30 years is 72 per cent, and the chances of a 7.0 event is 51 per cent—numbers high enough to raise concern, but not to provoke people into thinking about moving out.

Earthquake forecasts have a useful role to play in providing a general guide to roughly when the next quake might occur. For disaster managers and the emergency services, however, they are not particularly useful. Something much more precise is needed that will tell them, with sufficient warning, when and where the next earthquake will strike and, ideally, how big it is going to be. Something along the lines of: 'there is a 90 per cent certainty that a Moment magnitude 6.8+ earthquake will happen the Tuesday after next on the Hayward Fault, east of San Francisco'. Unfortunately, the reality is that we can't make such precise predictions, and this may, indeed, never be possible at all.

To be effective any earthquake prediction methodology must be shown to work *every* time. If it doesn't, then it is next to worthless. Claims have been made for the successful prediction of certain earthquakes, but they either don't hold up in the face of closer examination or are simply a consequence of a lucky break. Chinese scientists claim, for example, to have successfully provided a prediction before an earthquake close to the city of Haicheng in 1975. Following several months of the land surface tilting, water gushing from the ground, and the strange behaviour of animals inhabiting the area, more than 90,000 citizens were evacuated from the city on 4 February. The next day, just over 12 hours later, a magnitude 7 quake hit the city destroying over 90 per cent of the buildings, but taking virtually no lives because of the timely exodus. All the signs had suggested that a quake was on its way, but it seems likely that the scientists were simply fortunate in terms of telling the civil authorities when to evacuate. Certainly the method of prediction was not transferable to other earthquake-prone areas, and the Chinese failed utterly to warn citizens of the devastating Tangshan quake that occurred the following year.

All sorts of precursory signs occur in advance of large earthquakes, including measurable ground surface deformation, changes in water levels in wells, emissions of radon gas, increases in small earthquakes, changes in the electromagnetic properties of the bedrock, strange lights in the sky, and yes, unusual animal behaviour. The problem is that all these phenomena can and do occur without a major quake following them. For example, small earthquakes that presage a big quake may be interpreted as so-called *foreshocks*, but only in retrospect. Without a major quake following on, they are nothing more than small earthquakes. And, on many occasions, large quakes happen without any precursory smaller seismic events. Furthermore, how do you decide if your chickens or pigs are behaving oddly due to an imminent quake, or because of something else entirely?

Translating precursory signs into an accurate prediction of the timing and size of a coming quake is, then, beyond the realm of current earthquake science, and most seismologists now think that pursuing this particular prize is just not worth the effort. Much better to focus on ways and means to ensure buildings stay upright during a quake.

There is also evidence that accurate prediction might be less a holy grail and more a poisoned chalice. Consider what would happen, for example, if we could accurately predict a major earthquake many months or even years ahead. Imagine in 30 years' time that seismologists have developed a technique that allows them to pinpoint the timing and size of an earthquake two years ahead, with an accuracy of two or three weeks. Let's say that in early 2050, the USGS issues a warning that a magnitude 8.1 quake will strike Los Angeles in November 2052. The immediate effect would be devastating. Property prices would plummet and the real estate business would collapse as buyers for property in the region evaporated. Insurance companies would pull out just as quickly, leaving those with unwanted properties on their hands without cover. Major companies would make plans to move elsewhere, offloading their workforces and creating a huge unemployment problem. It is not hard to believe that, even before the earthquake struck, its prediction might have caused almost as much, if not more, damage to the economy of the region. Perhaps then it would be better if the holy grail of successful earthquake predicting—like the Holy Grail itself—remained an ethereal objective always just out of reach.

Earthquakes don't kill people: Buildings kill people

Earthquakes kill and destroy in many ways, including as a consequence of fires triggered by broken gas mains and naked flames; landslides; and, of course, tsunamis. But most of the destruction and three-quarters of the lives lost to earthquakes occur because of building collapse. It isn't surprising, then, that seismologists and earthquake engineers never tire of pointing out the seemingly obvious fact: *earthquakes don't kill people, buildings kill people.* What they are getting at is the simple point that if buildings were properly constructed in earthquake zones, so that they can stay upright when the ground shakes, then no-one need lose their life. Unfortunately, this goal remains a far distant achievement, as the massive levels of destruction and loss of life in recent quakes testify.

In 1999, a magnitude 7.4 earthquake struck the Izmit region of Turkey, obliterating 150,000 buildings and taking more than 17,000 lives. Many apartment blocks simply *pancaked*: successive floors collapsing to form a stack of concrete slabs beneath which opportunities for survival were minimal (Figure 6). The same thing happened again in the Turkey–Syria quakes of 2023. In January 2001, a severe earthquake shook the Bhuj region of Gujarat state in north-western India, flattening 400,000 homes and killing perhaps 100,000 people. Many of the deaths resulted

6. Pancake collapse of a school in Mexico City following the 2017 Puebla earthquake took the lives of more than 20 children.

from the traditional construction methods used in the region, which involved the building of homes with enormously thick walls made of great boulders held together loosely with mud or cement, beneath heavy stone roofs. When the ground started to shake, these buildings offered little resistance, collapsing readily to crush those inside. In 2003, a moderate earthquake in southern Iran took 26,000 lives in the city of Bam, as the traditional adobe (mud brick) buildings put up little or no resistance to the ground shaking. In 2015, most of the 9,000+ people whose lives were lost in the Nepal earthquake occupied poorly constructed 'informal' buildings, put together in an *ad hoc* manner, often on unsuitable terrain.

The close to 300,000 buildings destroyed or severely damaged in the 2010 Haiti earthquake were built poorly, without any thought of the effects strong seismic shaking would have on them. And the list goes on: poor construction standards were almost entirely responsible for an estimated 780,000 or more buildings destroyed or badly damaged during the 2005 Kashmir quake, and the five million buildings flattened in the 2008 Sichuan earthquake, together resulting in the loss of close to 180,000 lives.

Badly built schools were hit especially hard in both events. In Kashmir, 19,000 children were killed due to the widespread collapse of school buildings, while in Sichuan more than 7,000 school buildings failed, leading to huge loss of lives amongst the students. This sparked the so-called 'tofu schools' scandal, which saw local and national government blamed and chastised over the weak state of school buildings and their inability to remain standing during any significant level of ground shaking. Poorly built schools were once again hit hard during a moderate (magnitude 5.6) quake in western Java in 2022, leading to a disproportionate number of children being among the death toll of around 600.

The technology to make buildings *life-safe*—meaning that no-one inside need lose their life when a quake hits—exists and is well established. In developed nations and regions characterized by high seismic risk, it is used widely. In California, for example, where nearly everyone lives within 50 kilometres of an active fault, strict building codes ensure that new buildings can withstand the worst expected ground shaking without falling down. There are, however, huge numbers of older buildings that need to be retrofitted to make them safe, and this is a massively costly business. Some older properties have been made life-safe, but many others have not. Currently, it is estimated that there are more than 1.2 million homes in the state that are at risk of being made unlivable following strong seismic shaking. Worldwide, the cost of making all buildings safe in every seismic hotspot would be prodigious, and most countries simply don't have the cash or, in many cases, the expertise to do this.

Ensuring buildings stay upright during violent shaking is no easy business, primarily because a typical suite of seismic waves is capable of shaking a building in every direction. There are four main varieties of seismic wave, each of which shakes the crust it travels through in a different way (Figure 7). Following the triggering of a quake, the first seismic waves to arrive are the speedy *P-waves* (Push waves). These move by squeezing and stretching the rock through which they travel rather like a concertina being pulled and pushed, or like a row of wagons being shunted back and forth by a railway engine. Next to arrive are the *S-waves* (Shear waves), which cause the crust to move up and down or from side to side, perpendicular to the line of travel. Both *P-* and *S*-waves are known as *body waves*, as they travel through the body of the Earth, but there are others, known as surface waves which, as the name suggests, travel close to the surface. As these take a longer route, they arrive later than the body waves. There are two types of surface wave: *Love waves* involve horizontal movement perpendicular to the direction of travel, resulting in a gentle side-to-side swaying of the ground. *Rayleigh waves*, on the other hand, involve vertical movement, causing a water-wave-like rolling motion of the ground surface. Together, surface waves are especially damaging, setting tall buildings and bridges swaying,

Main types of seismic waves

7. Seismic waves comprise body waves (*S-* and *P*-waves), which travel through the solid Earth; and surface waves (Love and Rayleigh waves), which—as the name suggests—are propagated at the surface.

bringing down elevated highways, and snapping communication and power cables.

The directly damaging effects of seismic waves on buildings can be made worse by the nature of the underlying geology. In the 1964 Niigata (Japan) earthquake, for example, liquefaction of soft sediment caused entire apartment blocks to sink into the ground, tilt sharply, or fall backwards. Liquefaction also caused significant damage to the San Francisco Marina District in the 1989 Loma Prieta quake, and to the port of Kobe (Japan) during a major earthquake there in 1995.

Soft sediment can also significantly amplify ground motions, hugely increasing the capacity for destruction. Probably the best example of this occurred during the 1985 Mexico City earthquake. The capital was fully 350 kilometres from the epicentre of the quake, and damage across much of the city was relatively minor. The western part, however, was constructed on an old lake bed comprised of unconsolidated sediments that enormously augmented the degree of shaking, resulting in catastrophic damage.

Notwithstanding the underlying geology, however, earthquakes are immensely destructive mainly because most cities in regions of high seismic risk are dominated by buildings that are simply not put together well enough to withstand the severe ground shaking caused by a large quake. Whether or not you are likely to survive a serious earthquake depends these days—to a very large degree—on where you live. As for most hazards, if you are lucky enough to live in a developed nation, where stringent seismic building codes are in place, your chances of survival are far higher than if you don't. Many earthquake-prone countries in the Majority World do have building codes in place that are designed to minimize damage due to seismic shaking, but often these codes are inadequate if the size of the earthquake is greater than expected, or are simply not enforced. Undoubtedly, corruption plays a significant part, and as a result corners are cut. The consequences of this become all too apparent whenever a big quake strikes an inadequately prepared city, and this is most recently exemplified in the 2023 Turkey–Syria quakes. But this situation need not arise. Relatively cheap means are available that can significantly reduce the level of building collapse and loss of life, and there is no excuse for not ensuring—at the very least—that hospitals and schools are made life-safe.

In those parts of the Developed World where the threat of an earthquake is high, where there is the money, and where expertise is plentiful, all sorts of engineering solutions are available to ensure buildings can withstand powerful ground shaking. Internal shear walls can be added to strengthen poorly supported ground floor spaces such as car parks and garages; the connections between vertical columns and horizontal supporting beams can be upgraded to minimize collapse; external bracing can be added to reduce lateral movement; critical buildings, such as schools or hospitals, can even be fitted with giant shock absorbers beneath them to intercept the worst of the shaking.

In Majority World countries, improving building safety during an earthquake is often compromised by a combination of corruption, insufficient funding and expertise, a lack of political will, and a focus on more immediate priorities; but straightforward measures can still be taken to reduce the chances of building collapse. The best time to increase resilience to earthquakes and reduce future death tolls in nations of limited means is to strike when the iron is hot—in other words, immediately after a seismic disaster. This is when shock and grief are still fresh, memories haven't begun to fade, and day-to-day issues have not yet regained centre stage. The key to improving a building's response to serious ground shaking is to ensure that it is both strong and *ductile*—in other words flexible enough to 'give' or 'go with the flow'. Many buildings in Majority World countries are, however, constructed from unreinforced masonry (brick, adobe, stone, or concrete blocks that may or may not be bound together by mortar), which is neither strong nor ductile and, therefore, especially vulnerable even to relatively low levels of shaking. These can be made safer by ensuring that the walls and floors are securely attached so as to minimize the risk of pancaking, while additional vertical support can be provided by adding pillars of reinforced concrete. Because such retrofitting typically costs between 25 and 75 per cent of the cost of the building, however, it is often cheaper to rebuild—following a quake—using earthquake-resistant materials such as reinforced concrete, which only adds 5–10 per cent to the cost. If, as is likely, resources for retrofitting or constructing new buildings are limited, the focus must always be on ensuring that school and hospital buildings are as quake proof as possible, to keep children safe during any future earthquakes and to guarantee that facilities will be available to treat the injured.

It will take a great deal of time, however, for rebuilding and retrofitting initiatives to make any sort of dent in the number of vulnerable buildings in Majority World countries where the level of seismic hazard is high. Inevitably, other priorities get in the way, jostling for attention and the limited funding available. Consequently, it would be naïve to think that the end of major earthquake disasters is in sight. Indeed, some seismologists and earthquake engineers are already warning of the potential, in the near future, for a single large earthquake to take upwards of a million lives.

Tsunamis—the sting in the tail

Go back to the beginning of the century and few people knew what a tsunami was. Indeed, these devastating phenomena were still typically referred to as 'tidal waves', even though they have nothing to do with tides. By the last decade of the 20th century, tsunamis had taken a few thousand lives in Indonesia (1992) and Papua New Guinea (1998), and hundreds of lives had been lost to smaller tsunamis in Nicaragua, Japan, Java, and Turkey. None, however, had made it big on the world media stage or captured public attention in any meaningful way. This all changed at just before 8am on 26 December 2004, when a 1,600 kilometres long stretch of the submarine Sunda Megathrust ruptured, triggering a magnitude 9.1 earthquake, which in turn spawned a devastating tsunami that took the lives of people as far away as east Africa. The ruptured section of the mega-thrust stretches from northern Sumatra northwards almost as far as Burma, and marks the boundary between the subducting Indian Plate to the south and west, and the overriding Burma Plate to the north and east. It had been accumulating strain for centuries and was known to present a massive threat to the region. Nonetheless, no tsunami warning system was in place and there had been no official public education campaigns flagging what people should do in the event of a major quake likely to promote a tsunami. The result was carnage. The first waves—in excess of 50 metres high in places—reached Sumatra's northernmost province of Aceh within 20 minutes, flooding several kilometres inland. The level of destruction was near complete, and many tens of thousands of lives were snuffed out within minutes. In total, the tsunami caused more than 167,000 deaths in Indonesia.

In the absence of a regional warning system, countries further afield were also completely unprepared. In Sri Lanka, fully 1,700 kilometres from Sumatra, the tsunami took more than 35,000 lives, and a further 12,000 in India. In Thailand, the death toll was in excess of 8,000, while hundreds more lost their lives in the Maldives and along the east African coast, even though the waves took several hours to reach there. The total death toll—for quake and tsunami—was a shade under 230,000.

Suddenly everyone knew the meaning of 'tsunami' and its potential to provide a deadly sting in the tail of an earthquake. Because tourist destinations such as Thailand, Sri Lanka, and the Maldives were affected, thousands of visitors from Europe, North America, and Australia, who would normally have been insulated from such geophysical slaughter, also lost their lives, including more than 500 Swedish and 91 British citizens. Almost everyone in the UK seemed to know someone who had been affected or who knew someone else who had, so familiarity with the phenomenon expanded hugely.

And then, on 11 March 2011, it happened all over again, this time off the north-east coast of Japan, close to the city of Sendai in the prefecture of Tohoku. The source of this second devastating tsunami in just seven years was another magnitude 9.1 quake on a mega-thrust where the Pacific Plate subducts beneath northern Japan. The tsunami risk had been known, and a warning was sent out to the public just 30 seconds after the quake. Both the size of the earthquake and the scale of the tsunami were, however, significantly underestimated. As a consequence, coastal walls designed to provide protection from tsunamis in populated areas were too low, and were easily and rapidly overtopped. This, in turn, meant that more than 100 designated tsunami evacuation points were inundated. Waves up to 40 metres high reached 10 kilometres inland, causing destruction on a colossal scale and taking nearly 20,000 lives. Tsunamis triggered by earthquakes have continued to take lives during the years since the Tohoku event, including 4,000 or more following a quake off the Indonesian island of Sulawesi in 2018.

Anyone who has seen video footage of the 2004 tsunami and the helicopter shots of the sea flooding across north-east Japan in 2011 will appreciate the almost unbelievable power of water to kill and destroy. Every cubic metre of water weighs a tonne, and in a tsunami, trillions of tonnes of water crash onto a coastline faster than a human can run. Being caught in the open is like being hit by a truck, and the chances of survival are slim. Many lives are also lost in the surging, debris-laden, waters, as they drain back off the land.

The speed at which a tsunami travels is another reason for its huge destructive capacity (Figure 8). Because mega-thrust

8. The devastating aftermath of the 2011 Japan tsunami at Kesennuma in the prefecture of Miyagi in north-east Japan.

earthquakes happen not far offshore, the waves can reach the nearest land mass in as little as 10 minutes, often long before any warning has been issued. In deep water, tsunami velocities can be in excess of 800 kilometres an hour, so they can rapidly cross an ocean basin. While not obvious in deep water, as the seabed rises on approach to land, and the water depth reduces, so tsunamis build to heights that can exceed 50 metres at the coast.

It would be easy to imagine, on the basis of recent events, that potentially devastating tsunamis are a product only of the subduction that goes on around the Pacific Rim, and the mega-thrust earthquakes that are the result, but this is not the case. Quake-triggered tsunamis can happen almost anywhere. In 1755, an offshore earthquake generated a tsunami that contributed to the almost complete destruction of the Portuguese capital, Lisbon. Many tsunamis have also been recorded in the Mediterranean since around 500 BC, resulting in serious loss of life following the 1783 Calabria and 1906 Messina earthquakes, both in Italy. Even normally geologically quiet Canada is not immune. In 1929 a submarine earthquake off the Atlantic coast caused a submarine landslide that drove a tsunami onto Newfoundland, taking a couple of dozen lives and leaving 1,000 without homes.

While the two greatest tsunami catastrophes of modern times have resulted from great submarine earthquakes, tsunamis can form in other ways too, including due to volcanic explosions and collapses, landslides, asteroid impacts, and even the weather. In 1883, the destruction of the island volcano of Krakatau (Indonesia) led to the formation of a tsunami up to 40 metres high along the coasts of Java and Sumatra that was responsible for the majority of the 36,000 lives lost. In northern Italy in 1963, the collapse of a mountainside into a high dammed lake resulted in the displacement of 50 million tonnes of water onto the town of Longarone below. More a splash wave than a true tsunami, the destruction was on a par, the town being wiped from the map and 2,000 of its residents killed. The greatest known tsunami was that caused by the impact, 66 million years ago, of the $c.14$ kilometres wide asteroid that crashed to Earth off the coast of Mexico, and that finished off the dinosaurs, along with many other animals. The initial wall of water is estimated to have been 1.5 kilometres high, and still 10 metres high when the waves reached New Zealand 10,000 kilometres away. Within 48 hours of the impact, hardly a coastline on the planet would have been unaffected.

Tsunamis will keep happening, but in terms of their killing capacity, there is plenty we can do to spike their guns. Homes can be made safer by supporting first-floor living space on reinforced concrete pillars, allowing water to pass below unimpeded. The cost of doing this on any scale, particularly in those Majority World countries most at risk, would, however, be prohibitive. Far better to provide community refuges to the same plan, to upgrade tsunami warning networks, and build lines of communication between those issuing the warnings and the public under threat. All these measures need to work too, otherwise they are next to useless. For example, while the Indian Ocean does now have a tsunami monitoring network in place, it is not currently fully functional.

In ideal circumstances, members of the public under threat should be warned of a possible tsunami within a minute or so of a quake happening, via radio, sirens, or mobile phones—preferably all three. A public education campaign, along with evacuation exercises, should ensure that everyone knows what to do: head inland, uphill, or to one of the community refuges. In case of communication failures, coastal residents should also be trained not to wait for warnings, but to act as soon as they feel the long-duration ground shaking associated with a mega-thrust earthquake. We may not be able to stop tsunamis, but we can take measures to ensure the huge death tolls of 2004 and 2011 are not repeated.

Where next?

I am often asked where I think the next big earthquake or volcanic blast will happen, and if there is any consistency in my predictions it is that I am invariably wrong. This isn't to say that we don't have a clue as to where future geophysical flashpoints are, it's just that Nature doesn't always play ball. In the case of earthquakes, where the records are good enough to determine reasonably accurate return periods, we can make a decent fist of speculating on when the next quake might be expected, but even here the intervals between quakes might be sufficiently variable as to make the time envelope for the next event decades, or even centuries, long. Then there are what are called *seismic gaps*, which—in simple terms—mark segments of active faults that have not slipped for an unusually long time when compared to other segments. So what do these various pointers say about our world's seismic future? Here are a few places worth keeping an eye on.

Hardly surprisingly, Tokyo is very near the top of the list of at-risk cities—a real concern when you consider that the population of the metropolitan area is a staggering 37 million. The most recent study suggests that the probability of a major earthquake striking the Japanese capital in the next three decades could be as high as 70 per cent, and would cause huge loss of life and colossal damage. Another area of serious concern is an unruptured segment of the Sunda Megathrust that could fail at any time, resulting in a magnitude 8.5+ quake and accompanying tsunami close to the Sumatran city of Padang. Staying in South-East Asia, the cities of Manilla (Philippines), Jakarta (Indonesia), and Taipei (Taiwan) are all exposed to significant levels of seismic risk.

Heading north, South Asia has its fair share of major seismic threats too. Eastern Bangladesh sits above a mega-thrust segment, the rupture of which could produce a magnitude 9 earthquake. Further north, along the southern front of the Himalayas, a number of magnitude 8+ quakes are 'due' that could have a severe impact on the Indian capital, Delhi, and other major cities in the region.

I think I am on pretty safe ground if I forecast that California will experience a significant earthquake soon. After all, the state sees a magnitude 6+ quake happen every few years. Because of the stringent seismic building codes, however, damage tends to be minimal and loss of life rare. There are, however, bigger earthquakes on the horizon that could be expected to have far greater impacts. In the Los Angeles area, the USGS estimates that there is a significant likelihood of a magnitude 7.5+ quake happening in the next couple of decades. This could result in some 2,000 deaths, and destruction and damage totalling US$200 billion. Further north, the biggest worry in and around San Francisco is a future major earthquake on the Hayward Fault, which runs along the eastern edge of San Francisco Bay. Based on return periods of previous quakes, another 'big one' is due here, which the USGS has termed a 'tectonic time bomb'. Another time bomb may be ready to go off on an active fault system located beneath the US state of Missouri—thousands of kilometres from the nearest plate margin. As mentioned earlier, a series of quakes (the New Madrid earthquakes) rocked an area of hundreds of thousands of square kilometres in 1811 and 1812, some larger than magnitude 7. Then the region affected was sparsely populated, but now it hosts major urban centres, including Memphis and St Louis, none of which are prepared for a big earthquake and its consequences. More than 4,000 small quakes have been recorded here since 1974 and the possibility of a future major earthquake in what is known as the *New Madrid Seismic Zone* is being taken very seriously. One study suggests that the likelihood of a magnitude 6+ quake happening in the next half century is as high as 40 per cent. Meanwhile, FEMA (the US Federal Emergency Management Agency) has warned that a future quake here could cause widespread damage across eight states and result in the highest economic losses of any US natural disaster to date.

Head still further north and you come to one of the biggest seismic threats in the developed world. Running from northern California all the way up to the Canadian province of British Columbia is the 1,000-kilometre-long submarine Cascadia Megathrust, marking the subduction zone along which the Juan de Fuca plate to the west is plunging down beneath North America. In 1700, a magnitude *c.*9 earthquake triggered a Pacific-wide tsunami, and the next such event is awaited with trepidation. The probability of a repeat performance is estimated at between 5 and 10 per cent in the next 30 years, which are not especially worrying odds, but certainly high enough to think upon during sleepless nights. The chance of just the southern segment of the mega-thrust rupturing, to produce a magnitude 8–8.5 quake may be as high as 30 per cent over the same period. A combination of the destruction from such a quake itself and the ensuing tsunami is forecast to take at least 13,000 lives.

Elsewhere on the planet, the Caribbean is a place to watch. The region is crossed by numerous active faults and also hosts a small subduction zone above which the volcanoes of the Lesser Antilles are located. Typically, earthquakes are accompanied by tsunamis, including two lethal quakes that struck the Kingston area of Jamaica in 1692 and 1907. Other major quakes took many lives in Guadeloupe in 1843 and the Dominican Republic in 1946. Most recently, the catastrophic 2010 Haiti quake, and the follow-up in 2021, highlighted just how vulnerable population centres in the Caribbean are to seismic shocks. Future concerns are focused on a big earthquake and accompanying tsunami triggered on one of the faults to the north of Puerto Rico, or in the Lesser Antilles.

Last, but not least, there is Istanbul; a city of more than 15 million people—20 per cent of Turkey's population—sitting just to the north of the unruptured western end of the North Anatolian Fault. When the segment immediately to the east ruptured in 1999 at Izmit, driving a magnitude 7.6 earthquake, the devastation was terrible, and more than 17,000 lives lost. Understandably, then, the inhabitants of Istanbul await the next quake with justifiable concern, particularly in light of the carnage caused by the 2023 earthquakes in the south of the country. A magnitude 7+ earthquake is expected to strike Istanbul within a decade, and there can be little doubt that this will cause hundreds of thousands of buildings to collapse or suffer serious damage, costing many thousands of lives.

Chapter 3
The volcanic menace

Molten planet

Inside our world, prodigious quantities of heat arising from the breakdown of radioactive isotopes are constantly heading for the surface, while at the same time heat left over from the planet's violent genesis is also still making its way outwards. The total annual flow of heat from the Earth's interior is a colossal 47 Terawatts (or 47,000 billion watts)—more than two and a half times the amount of energy used by human civilization over the same period. Heat is continually dissipating at a low level across every square metre of the planet's surface, but it is also expelled far more spectacularly in the form of eruptions from the world's active volcanoes. The magma that feeds these has its origin in the *asthenosphere*, the partially molten layer beneath the brittle lithosphere that makes up the tectonic plates, where the pressure and temperature conditions are just right for rock to become molten and stay that way. Above this, temperatures are too low for melting to happen; and at greater depths, increasing pressures prevent melting.

Since 1800, around 530 volcanoes have erupted, but quite a few that have stayed dormant over the last couple of centuries may just be biding their time, so the total number of potentially active volcanoes is likely to be far higher. In fact, more than 2,000 volcanoes have erupted in the last 10,000 years or so, any one of which could burst into life again. Across the last decade, an average of a little more than 70 volcanoes have been in eruption every year, some continuously, others for the first time in decades or even centuries. Communities have learned to live with those volcanoes that are constantly rumbling, such as Stromboli (Italy), Kilauea (Hawaii), and Sakurajima (Japan), so the greatest threat comes from volcanoes that have not erupted in living memory or, in some cases, far longer. Not least, this is because volcanoes that have long been dormant tend to wake with extreme violence and to lethal effect.

While being nowhere near on a par with earthquakes in terms of levels of destruction and lives lost, the world's volcanoes are a clear and present danger to the more than half a billion people that live in their shadows. Individual eruptions have taken more than 20,000 lives at a go, and impacted directly the lives of millions more. Furthermore, the physical consequences of the biggest volcanic blasts can be felt globally, a scale beyond the reach of even the biggest earthquake.

Like earthquakes, nearly all of the world's volcanoes are sited at plate boundaries. Around two-thirds, maybe 900 in all, sit above the circum-Pacific subduction zones that define the appropriately named *Ring of Fire*. In addition, subduction-related volcanism can be found in the Caribbean and the Mediterranean. Magma also reaches the surface along the constructive plate margins that are represented by the system of Mid-Ocean Ridges. Much of this feeds submarine volcanoes, but some cools and accumulates over time to build volcanic islands. These include Ascension, St Helena, and Tristan da Cunha, which sit on the Mid-Atlantic Ridge and, of course, Iceland—an especially dynamic volcanic hotspot that hosted 39 eruptions in the 20th century alone (Figure 9). Exactly why Iceland is such a major focus for volcanic activity—compared to other points on the Mid-Atlantic Ridge—is not certain, but the consensus is that it sits above a rising plume of

9. The ash plume of Iceland's Eyjafjallajökull volcano wrought chaos across the flight paths of Europe in Spring 2010.

hot mantle material that may have its source deep within the planet, at the core–mantle boundary. Such so-called *mantle plumes* also reside beneath the Hawaiian Island volcanoes, the Azores, Cape Verde Islands, and the infamous Yellowstone super-volcano, from all of which any plate boundaries are remote. Here and there, volcanoes have become established where local geological circumstances have resulted in points of weakness that have been exploited by rising magma. These include Mount Etna in Sicily and Vesuvius, overlooking the Bay of Naples. Elsewhere, in east Africa, volcanic activity is associated with *rifting* (splitting apart) of the African Plate and the formation of a new constructive plate margin. Here, eruptions of the Nyiragongo (DRC) volcano cost lives in 1977, 2002, and 2021.

As with earthquakes, the deadliest and most destructive volcanoes squat—broadly speaking—above subduction zones, so many population centres around the Pacific Rim and in the Caribbean have to put up with both. Here, typically at depths of more than

100 kilometres, fluids from the subducting plate rise into the mantle, reducing its melting point, and causing it to 'sweat' *basalt* magma. This has a low viscosity but, as it rises due to its buoyancy, it begins to crystallize out minerals that are low in *silica* (SiO_2), leaving the liquid part of the magma increasingly silica rich. In this way, basalt magma can evolve over time to the progressively more silica-rich varieties—*Andesite*, *Dacite*, and *Rhyolite*. All magmas, even those of basaltic composition, contain a substantial amount of silica and, to a large degree, it is the silica content of magma that controls its viscosity, so that a higher silica magma will be stickier than one with a lower silica content. Because any contained gases find it harder to escape from sticky magma, silica-rich magmas tend to feed violently explosive eruptions that are characterized by pyroclastic flows and surges and great columns of volcanic ash.

Major eruptions at subduction zone volcanoes in the last three decades or so have happened at Pinatubo (Philippines) in 1991; Volcán Hudson in 1991 and Cordón-Caulle in 2011–12 (both Chile); and Hunga Tonga-Hunga Ha'apai (Tonga) in 2022. So far this century, lives have been lost during eruptions of close to 20 subduction zone volcanoes, including Krakatau, Merapi, Semeru, and Sinabung (Indonesia); Ta'al (Philippines); White Island (New Zealand); Volcán de Fuego (Guatemala); and Ontake (Japan).

In sharp contrast to subduction zones, basalt magma formed by mantle melting beneath the Mid-Ocean Ridge system and at most mantle hotspots is formed at shallower depths and so reaches the surface before it has had a chance to undergo significant crystallization. Consequently, it remains relatively silica-poor and has a lower viscosity, feeding less violent, so-called *effusive* eruptions that are characterized by spectacular lava fountains and quietly oozing lava flows. Occasionally, eruptions can be explosive, and are more widely disruptive as a result.

Every shape and size

Volcanoes come in a wide variety of shapes and sizes. Topography ranges from the iconic Mount Fuji (Japan) cone, a so-called *stratovolcano* built up of layers—or strata—of ash and lava, to gently sloping *shield* volcanoes, constructed almost entirely from lava flows, such as Hawaii's Mauna Loa. Sometimes, as at Yellowstone (Wyoming, USA) and Campi Flegrei (Bay of Naples, Italy), a volcano doesn't form any sort of topographic high and can be difficult to spot. Only detailed geological investigation will reveal the form and structure of these low-lying giant craters, known as *calderas*, which represent the remnants of volcanoes that have collapsed or blown themselves apart.

Volcano form is closely related to the chemistry of the magma, with stickier, more silica-rich magmas forming cones, mainly comprising ash fall or pyroclastic flow deposits, or steep-sided domes of viscous lava; while runnier, relatively silica-poor magmas build shallow shields. Magma chemistry is also a major determinant of the so-called *style* of eruption, which ranges from the quiet effusion of lava to colossal explosions.

A number of eruption styles are recognized, most named after volcanoes that have been observed to commonly display them. A *Hawaiian* style, for example, typifies activity at the Hawaiian volcanoes, involving the relatively gentle extrusion of very fluid lava with a low gas content. Eruption often occurs from fissures along which independent small cones build as a consequence of focused lava fountaining, a phenomenon that is self-explanatory. A *Strombolian* eruptive style—named after the island volcano, Stromboli, north of Sicily—is a tad more vigorous, sending up small firework-like displays every few minutes, driven by bursting gas bubbles in the rising magma. Far more violent are the *Vulcanian* explosions that have characterized activity at Vulcano, another small island volcano in the Mediterranean. These involve the shattering and expulsion of a lava cap or dome by high pressure gases unable to escape easily from the magma, which is stickier than that feeding Hawaiian or Strombolian eruptions.

There are a number of other styles but not enough space to introduce them all here. One final style that must be mentioned, however, is the *Plinian.* This involves extremely explosive eruptions that blast out huge columns of ash and gas to altitudes as great as tens of kilometres. Instead of being named after a specific volcano, the term *Plinian* acknowledges the Roman natural philosopher, Pliny the Younger, who observed the 79 AD eruption of Vesuvius that killed his uncle (Pliny the Elder) and buried the city of Pompeii. Although ominously quiet since 1944, many previous eruptions of Vesuvius, including Pompeii's nemesis, have displayed this style of activity.

A key point about eruptive styles is that a single volcano may display several of them over the course of a number of eruptions, or even during a single eruption. Where a volcano reawakens gently, for example, perhaps with small explosions or the slow growth of a lava dome, this can be—and often is—followed by massive blasts that trigger the release of pyroclastic flows and a column of ash that reaches the stratosphere.

Volcanic eruptions also come in a range of sizes, and a number of scales have been devised to allow the sizes of volcanic events to be compared. One of the most commonly quoted is the *Volcanic Explosivity Index* or VEI (Table 3), devised in 1982 by volcanologists Chris Newhall and Steve Self, primarily to allow estimation and comparison of the magnitudes and intensities of historical eruptions. *Eruption magnitude* refers to the mass of material erupted, while *eruption intensity* is a measure of the rate at which material is expelled. The index is logarithmic (like the earthquake magnitude scales) which means that each point on the scale represents an eruption 10 times larger than the one immediately below. Thus a VEI 5 is 10 times larger than a 4;

Table 3. The Volcanic Explosivity Index

VEI	Ejecta volume (km³)	Description	Eruption column (height in km)	Stratospheric injection
0	<0.0001	Effusive	0.1	None
1	0.0001	Gentle	0.1–1	None
2	0.001	Explosive	1–5	None
3	0.01	Severe	3–15	Possible
4	0.1	Catastrophic	>10	Definite
5	1	Cataclysmic	>10	Significant
6	10	Colossal	>20	Substantial
7	100	Super-colossal	>20	Substantial
8	1,000	Mega-colossal	>20	Vast

Notes: The VEI classifies eruptions on the basis of the volume of ejected material, the height of the eruption plume, and whether or not there is injection of material (most notably sulphur gases that can affect the climate) into the stratosphere.

a VEI 6 is 100 times larger; and a VEI 7 1,000 times larger. At the bottom of the index, the gentle effusions of lava that characterize most eruptions of Kilauea and Mauna Loa on Hawaii score a measly 0, while mildly explosive eruptions that release sufficient ash to perhaps cover London or New York in a light dusting would register at 1 or 2. To most volcanologists, however, things don't really start to get exciting until higher values are reached. VEI 3 and 4 eruptions are described, respectively, as 'moderate' and 'large'. This translates into blasts big enough to cause local devastation, sending columns of ash up to 20 kilometres into the atmosphere and burying the surrounding landscape under piles of volcanic debris a metre or more deep.

In 1994, the town of Rabaul in New Britain (Papua New Guinea) was destroyed by a VEI 4 eruption, and a few years later, in 1997,

Plymouth, the capital of the Caribbean island of Montserrat, suffered the same fate. Eruptions that score a 5 on the scale, such as the intensively studied 1980 blast of Mount St Helens (Washington State, USA) typically cause mayhem on a regional scale, while VEI 6 eruptions can be regionally devastating and the effects long-lasting. The 1991 Pinatubo eruption in the Philippines was the second largest eruption of the 20th century, ejecting sufficient ash and debris to bury central London to the depth of a kilometre and making hundreds of thousands homeless.

For the last VEI 7 eruption we have to go back more than two centuries to 1815, the year of the battle of Waterloo. As the armies of Wellington and Napoleon jockeyed for position across Europe, on the distant Indonesian island of Sumbawa, the long-dormant Tambora volcano ripped itself apart in a gargantuan eruption that may have been the largest since the end of the Ice Age *c*.10,000 years ago. Sir Stamford Raffles, then British lieutenant governor of Java, reported a series of titanic detonations loud enough to be heard in Sumatra 1,600 kilometres away. When the eruption ended after 34 days, it left behind 12,000 dead. In the ensuing months, however, a further 60,000 or so Indonesians succumbed to famine and disease as they struggled to find food and uncontaminated water across the ash-ravaged landscape.

Although the VEI is open-ended, nothing bigger than a VEI 8 has been identified, and an event on this scale has never happened during historic times. VEI 8 blasts have also come to be known as *super-eruptions*, and the volcanoes that host them as *super-volcanoes*. These prodigious explosions really are in a different league to the common-or-garden volcanic blast, ejecting at least 1,000 cubic *kilometres* of ash and debris. Not only are they capable of unimaginable destruction on the ground, but the vast volumes of sulphur gases they pump into the stratosphere can have a severe impact on the planet's weather and climate. You will read more about these volcanic blockbusters later and in the final chapter.

How volcanoes kill

The considerable variety of hazardous and deadly phenomena that originate at volcanoes means that they can kill and destroy in all sorts of different ways, some close up and personal, others remote or one step removed. The two principal products of volcanoes are lava and ash. Lava is the primary hazard at basaltic volcanoes that extrude relatively silica-poor, low-viscosity lava in the form of flows that can extend for tens or, occasionally, hundreds of kilometres from the eruption site. Even low-viscosity lava is strong enough to push buildings over and bury them. Having a temperature not far below 1,000°C, it will also ignite many materials and can quite easily start forest or bush fires. Generally speaking, lava flows move far too slowly to be a threat to human life, except in special circumstances. At the DRC volcano, Nyiragongo, eruptions feed especially fluid lava flows that are speeded up by gravity as they travel down steep slopes from an elevated lava lake. As a result, they move faster than a person or animal can run. In its 1977 eruption, lavas moving at more than 60 kilometres an hour took around 600 lives in just 30 minutes.

Volcanic ash is the highly fragmented magma formed during its explosive ejection from a vent. Strictly speaking, the term is used for particles that are less than 2 millimetres in size, but it often becomes a catch-all for any finely fragmented material ejected into the sky from a volcanic vent, including *lapilli* (literally 'little stones'). Larger ejected fragments include blocks of solidified lava from the walls of the eruptive conduit and so-called *bombs*. These are gobbets of liquid magma that are shaped as they travel through the air, taking on bomb-like forms that taper at both ends. Large blocks and bombs rarely travel more than a few kilometres from an active crater, and are therefore only a threat to communities very close to a volcano, and to especially gung-ho volcanologists and those caught by surprise.

During the biggest explosive blasts, what is reported as ash often turns out to be *pumice*, the foamy-looking low-density rock used during ablutions to abrade hard skin. Both pumice and ash can be erupted in immense quantities, burying buildings, destroying crops, and polluting water supplies and fisheries across a huge area. Heavy ash or pumice fall can reduce visibility almost to zero, making any travel impossible, and as little as 30 centimetres of wet ash can collapse the roofs of buildings. Metres-thick deposits can bring transport to a halt, and damage communications and power infrastructure. Heavy pumice and ash fall caused widespread roof collapse during the 1991 eruption of Pinatubo (Philippines), and most of the more than 800 deaths. Ash and pumice also buried much of the town of Rabaul (New Britain, Papua New Guinea) during an eruption of the local volcano in 1994. The long-term inhalation of fine ash can lead to serious health problems, including the lung disease, *silicosis*. The finer component of volcanic ash can easily travel for many thousands of kilometres, shutting down aviation, causing travel problems on the ground, infiltrating electronics, clogging filters, contaminating water and vegetation, and causing problems for those with respiratory conditions. In 1980, the VEI 5 eruption of Mount St Helens (Washington State, USA) dumped ash across almost one-fifth of the country, the clean up across Washington State alone costing millions and taking several months.

The most lethal and destructive of all volcanic hazards are the pyroclastic flows (from the Greek for 'fire' and 'broken in pieces')—ground-hugging clouds of high-temperature gases, ash, pumice, and sometimes blocks as big as houses—that career down the slopes of a volcano at speeds averaging 100 kilometres an hour, but often much faster. A more dilute form of pyroclastic flow, in which the gas component dominates, is known as a pyroclastic *surge*. These can be equally damaging and just as deadly to humans as true flows. The chances of surviving after being caught by a pyroclastic flow or surge are next to zero, and their destructive capacity is colossal. During the 1902 eruption of Mont

10. The ruins of St Pierre (Martinique) after the 1902 eruption; only two inhabitants in the town itself survived the onslaught of pyroclastic surges from the Mont Pelée volcano.

Pelée (Martinique) in the Caribbean, pyroclastic surges obliterated the capital, Saint Pierre, in minutes, leaving just two survivors from a population of around 28,000. Photos of the aftermath of this greatest volcanic disaster of the 20th century reveal a level of destruction comparable with that at the Japanese cities of Hiroshima and Nagasaki following the detonation of nuclear bombs (Figure 10).

Most pyroclastic flows are formed either from the collapse or explosive destruction of a viscous lava dome or from the gravitational collapse of an eruption column of ash and pumice. Both are equally devastating, although the former tends to contain large blocks of the disrupted dome, while the latter is more pumice dominated. Often pyroclastic flows travel for just a few kilometres, presenting a threat to those communities on or close to a volcano's flanks. Larger eruptions, however, can drive the flows to distances of tens of kilometres, and VEI 8 super-eruptions can generate pyroclastic flows capable of travelling more than 100 kilometres and covering many thousands of square kilometres.

Pyroclastic surges tend to be less confined by topography than the denser pyroclastic flows, meaning it is difficult to predict paths of travel. During the 1991 eruption of Unzen (Japan), a surge killed dozens of journalists, alongside the French volcano film-maker couple, Maurice and Katja Krafft, and US volcanologist Harry Glicken. Both pyroclastic surges and flows have taken lives during a number of eruptions in the last few decades, including at the Soufriere Hills volcano (Montserrat, Caribbean) in 1997, Ontake (Japan) in 2014, Merapi (Indonesia) also in 2014, and Volcán de Fuego (Guatemala) in 2018.

As well as flows of lava and hot ash and gas, torrents of mud also present a major hazard at some volcanoes. Volcanic mudflows—also referred to as debris flows—are known by the Indonesian term *lahars*, and can be formed in any one of a number of ways. These include heavy rain falling on slopes covered by unconsolidated ash or pumice, hot erupted material falling on or flowing across ice and snow fields, and the breaching or overtopping of a crater lake. Lahars can be hot or cold, depending on their mode of formation, but all are equally deadly. In 1985, at Colombia's Nevado del Ruiz volcano, pyroclastic flows melted huge quantities of ice at the volcano's summit, triggering lahars that buried the town of Armero, fully 50 kilometres distant, and took 23,000 lives here and at other affected communities (Figure 11). Following the VEI 6 eruption of Mount Pinatubo in 1991, lahars formed by torrential rains mobilizing pyroclastic flow deposits were still causing huge problems up to a decade later, repeatedly clogging river channels, which resulted in the flooding of major cities.

Often characterized by steep slopes, sometimes bulging with fresh magma, shaken by earthquakes, and perhaps rotted internally by volcanic gases, volcanoes are not the most stable landscape features. It should come as no surprise, then, that the flanks of many fail during their lifetimes, producing giant landslides. The most famous, and certainly best photographed, is that formed by

11. Aerial view of the remains of the Colombian town of Armero, buried under lahars from the Nevado del Ruiz volcano in 1985.

the collapse of Mount St Helens (Washington State) on 18 May 1980. This huge landslide—at nearly 3 cubic kilometres, the largest in modern times —detached a topographic bulge formed by rising magma which, unable to escape at the blocked summit crater, forced its way into the north flank. By removing the 'lid' on the magma, the slide also promoted a major (VEI 5) blast, which took 67 lives and deposited ash across 11 states.

The Mount St Helens slide pales into insignificance when compared with landslides that happened in pre-history at the biggest ocean island volcanoes. Submarine imagery has revealed debris from slides with volumes more than 300 times greater around the Hawaiian Islands, Canary Islands, and other volcanic archipelagoes. These collapses may have sourced *mega-tsunamis* with enough energy to cross entire ocean basins, about which, more in the final chapter. Nearer to our time, and at a far smaller scale, a volcanic landslide at Unzen volcano (Japan) in 1792 triggered a tsunami that took more than 14,000 lives.

A little more than a century on, in 1888, a tsunami spawned by another slide, this time at Ritter Island (Papua New Guinea), killed more than 3,000.

Marine volcanoes can also trigger tsunamis in other ways, such as violent explosions, foundering (vertical collapse), and the wholesale entry of debris or pyroclastic flows into the ocean. The greatest death toll from a volcanic tsunami occurred during 1883, when Indonesia's Krakatau volcano was obliterated by a VEI 6 eruption. The waves reached the adjacent coasts of Java and Sumatra within 30 minutes, taking tens of thousands of lives. The disaster was re-enacted, albeit on a smaller scale, in 2018, when a tsunami arising from the collapse of Anak Krakatau (Child of Krakatau), which had formed in the 1883 caldera and emerged from the sea in 1927, killed more than 400 people in western Java and southern Sumatra.

As previously mentioned, it is possible that collapse-triggered mega-tsunamis can extend a volcano's perilous reach across ocean basins. Beyond this, however, the biggest volcanic eruptions can have a global impact, as proved to be the case following the 1815 eruption of Tambora. The c.200 million tonnes of sulphur-rich gases lofted into the stratosphere by the eruption mixed with water vapour to form a mist of sulphur aerosols that spread out across the planet, blocking incoming solar radiation. As a consequence, the following year—widely known as the *Year without a Summer*—saw the global average temperature fall by 1°C. Summer 1816 was the coldest in the northern hemisphere in more than 600 years, bringing snow, killing frosts, and cold, wet, conditions that devastated harvests in Europe and across eastern North America. The cold continued for another two years, causing famine and food shortages that led to widespread civil unrest during what is now known as the 'last, great, subsistence crisis in the western world'. So dreadful was the weather that it seems to have set just the right mood for Mary Shelley's vivid imagination to spawn its most famous offspring, *Frankenstein*, while the spectacular ash- and gas-laden sunsets are said to have inspired some of J. M. W. Turner's most brilliant works. Further afield, the climatic impact of the Tambora blast is also charged with disruption of the South Asia monsoon, causing harvest failures and widespread famine in that part of the world.

Another climate-altering volcanic event took place three decades earlier, this time in Iceland. The VEI 4 1783 Laki (or Lakagígar) eruption was very different from Tambora, pumping out lava rather than ash and pumice—around 14 cubic kilometres in all. The real problem, however, was the huge quantity of noxious gases, especially sulphur dioxide and fluorine, emitted alongside the lavas. Crops were wiped out, while ingestion of fluorine-contaminated vegetation killed off half the island's livestock, leading to a famine that reduced Iceland's population by a quarter. Further afield, a sulphurous smog settled over Europe, causing a sweltering summer and respiratory health problems that may have taken more than 20,000 lives in the UK alone. The eruption also brought an exceptionally severe 1783–4 winter in Europe, and one of the coldest winters on record in the United States. Here there are even reports of ice floes in the Gulf of Mexico. Droughts and famine as far afield as Japan, North Africa, and South Asia have also been blamed on the eruption.

Living safely with volcanoes

There are plenty of reasons why so many people live close to, or even on the flanks of, active volcanoes, not least the fact that volcanic soils are fertile and productive. On top of this, their elevation allows for a cooler and more pleasant climate in the tropics, where most are located, while the scenery, the views—and sometimes the lava—are great tourist selling points.

Since the 1783 Laki eruption, an estimated 220,000 people have died due to the proximal effects of volcanic eruptions, which works out at one-twentieth of the annual number of casualties due to earthquakes. Nonetheless, the aforementioned catastrophes on Martinique in 1902 and in Colombia in 1985 make it clear that living in the shadow of an active volcano is perilous, and can result in major loss of life. In addition, more than half of the volcanoes expected to erupt in the future will have been dormant for over a century, and have the potential to awaken with particular violence. In fact, 75 per cent of VEI 5 and 90 per cent of VEI 6+ eruptions have occurred at volcanoes that had been dormant for more than 100 years.

Keeping people safe on volcanoes is strongly dependent on a combination of reliable monitoring, accurate eruption prediction, and effective communication. Unlike earthquakes, the timing of a volcanic eruption can be predicted with reasonable accuracy. This is because the signs caused by magma rising towards the surface are obvious and detectable. As magma needs to break rock as it makes its way upwards through the crust, small earthquakes are generated that can be picked up by seismometers. Rising magma also needs to make space for itself, causing swelling of the surface above. This can be detected using a wide range of methodologies available for monitoring vertical and horizontal movements and slope changes, including GPS (global positioning system), precise levelling, electronic distance measurement, tiltmeters, and satellite radar measurements.

Acceleration in the numbers of small earthquakes and in the rate of ground swelling can be used to make predictions for when magma will reach the surface, and this has been applied to eruptions at a number of volcanoes, including Rabaul (Papua New Guinea), Soufriere Hills (Montserrat), El Hierro (Canary Islands), and Kilauea (Hawaii). Predictions can only be made, of course, if the seismic or ground deformation data are available, which means that a volcano needs to be monitored. Unfortunately, volcanologists are keeping an eye on less than half of the world's volcanoes. Worse still, many volcanoes present a serious threat in Majority World nations that have neither the cash nor the expertise to install effective monitoring networks. For this reason, useful predictions depend on the rapid establishment of emergency networks at the first sign of so-called volcanic *unrest*. This can take the form of faint ground tremors, sulphurous odours, the occurrence of steaming ground, or the increasing temperature of springs.

Because it takes time to build up a picture of any seismicity or ground deformation, an eruption prediction based on analysis derived from an emergency network may only be able to provide a few days' warning, and sometimes just a few hours. For this reason, the effective communication of the warning to local communities is critical if lives are to be saved. For this to happen, an established plan needs to be in place prior to unrest, which will allow decision makers to rapidly contact everyone under threat and tell them what to do. In advance of this, local populations should have been thoroughly educated about the potential threat, informed about what they should do if an eruption seems imminent, and, ideally, have undertaken practice evacuation exercises. Such forward planning can work wonders. At Rabaul (Papua New Guinea), residents had been fully educated, during a period of unrest in the mid-1980s, about the volcanic threat presented by the twin volcanoes close to the city. While no eruption followed, a big eruption did begin in 1994 following just 27 hours of rumbling. When the local volcano observatory called for evacuation, most of the population, recalling the lessons they had learned ten years earlier, had already gone.

Where there is sufficient time between the recording of unrest and the onset of an eruption for an adequate seismic and ground deformation dataset to be acquired, a system of pre-established alert levels can be used, linking escalating pre-eruptive activity, and ever more tightly constrained predictions of the start of the coming eruption, to responses required by affected communities. In ideal circumstances, this should seek to ensure that evacuation is undertaken in good time so that no lives are lost. In real life, however, things rarely work perfectly. Communication systems can break down, people can refuse to leave, and issues like traffic congestion and breakdowns can turn an orderly exodus into mayhem. In addition, it is perfectly possible that an episode of growing unrest may die down again without eruption—as happened at Rabaul in the mid-80s. Particularly if it happens post-evacuation, this can lead to a failure of trust in the monitoring scientists that makes it more difficult to get people to respond appropriately next time the volcano looks like waking up.

It is worth making the point that, although predicting when an eruption will start is now possible, that is pretty much it. Volcanologists can say little about what form the eruption will take, how big it will be, or how long it will last, except based on what they can glean from records, if there are any, of past activity, and geological mapping that reveals the products (e.g. lava, ash, pyroclastic flow deposits, etc.) of previous eruptions.

Individual eruptions can be over in days, or they can last for months or even years. In many ways, a moderate eruption that continues for a very long time can be, ultimately, more disruptive than a much larger one that is over in a week. The Soufriere Hills eruption on Montserrat, for example, began in 1995 and continued for 18 years, ending in 2013. The unusually long duration of the eruption brought immense suffering to the islanders, and led to the abandonment of dozens of communities.

Accurately predicting the size of an eruption is not possible either, and knowledge of previous activity doesn't necessarily help. Even if earlier eruptions have all been VEI 4s, the next one could still be a VEI 5, with all this would entail for management and mitigation of the event, not least the delineating of safe and no-go zones. The course of an eruption can also vary hugely from one volcano to another, and between different eruptions at the same volcano. In particular, the eruption climax, when most destruction occurs and most lives are lost, can come at any time. During the 1815

Tambora eruption, the climactic blast only came three years after the volcano woke from a long period of slumber. In contrast, the climax of the deadly 1902 Martinique (Caribbean) eruption, happened after just two weeks of activity. This is why timely evacuation is critical, and why extreme caution needs to be exercised in deciding when to allow people back.

Without any doubt there will be potentially devastating eruptions in the years and decades to come. Mount St Helens (VEI 5) events happen—on average—every decade or so, and there are several Pinatubo-scale (VEI 6) blasts every century. VEI 7 eruptions that can have a notable climate impact happen several times every millennium. It is even possible that global heating could drive an increase in eruptive activity. This has certainly been observed during past episodes of rapid climate change, especially where volcanoes lose their ice cover, and in marine settings where rising sea levels can modify the strain patterns within and beneath island and coastal volcanoes.

The number of people living close to an active volcano is certain to climb, for the reasons mentioned earlier and as a reflection of continued population growth. If future disasters on the scale of St Pierre and Armero are to be averted, there needs to be increased monitoring of those volcanoes that present the greatest threat, especially in the Majority World. Improved prediction methodologies could also play a key role in reducing volcanic risk, especially if they can say more about the likely size and duration of a coming eruption, and—particularly—if they can narrow down the timing of eruption climax.

The next big bang

A volcanologist taking a punt on the location of the next notable volcanic eruption is on a hiding to nothing. It is, however, a little easier than trying to guess where the next major earthquake will strike. This is because, as touched on earlier, volcanoes always display precursory signs that an eruption *might* be on the way. Again, as mentioned earlier, volcano unrest doesn't always, however, result in an eruption, or not imminently at least, and perhaps not for many years. As I write this, in June 2023, Askja and Grimsvotn (both Iceland), Ubinas (Peru), Lokon-Empung (Indonesia), and Kanlaon (Philippines) are all restless, and any or all of these volcanoes could be the source of the next high-profile eruption—or none of them.

There are a few other candidates worth considering too—volcanoes whose eruptions have return periods reasonably consistent enough for us to speculate that an eruption might be due. One of these is Teide, on the Canary Island of Tenerife, which erupted in 1705, 1798, and 1909 and showed some signs of restlessness in the first decade of the 21st century. Another is Mount Hood in the Cascade Range (Oregon, USA), which is located just 80 kilometres south of the city of Portland. Hood erupted in 1781 and in 1869, but has been dormant ever since.

A volcano that is always near the top of my list of those to watch is Vesuvius. Having been in almost continuous eruption in the preceding 300 years, the volcano has been ominously quiet since 1944. It might also be worth putting money on its near neighbour, Campi Flegrei, sited on the edge of the Bay of Naples. This giant caldera volcano has had some colossal eruptions in the pre-historic past, and has not erupted since 1538. It is always restless, but a significant and steady uplift over the last couple of decades is worrying, and research published in 2023 warns that the crust in the vicinity of the volcano is now stretched almost to breaking point.

Chapter 4
Storm force

The power of wind

Like earthquakes and volcanoes, windstorms have always been with us. Now, however, they are being supercharged by global heating so that their potential for destruction and loss of life is progressively escalating. At the most basic level, windstorms involve the anomalously rapid movement of batches of air in the lower atmosphere, or *troposphere*. They can be spatially limited, super-intense and short-lived, like tornadoes or, as is typically the case for tropical cyclones, they can cover vast areas and last for days, even weeks. The movement of the atmosphere that results in wind is a function of the Earth's rotation and the temperature difference between the tropics and the poles. Ultimately, the energy that drives windstorms comes from the Sun, but closer to home there are more specific causes. Tornadoes, for example, are spawned by the powerful air currents that accompany major thunderstorms, while tropical cyclones depend, for their formation and development, on the heat from warm, tropical oceans. At higher latitudes, many windstorms develop in situations where there are sharp gradients in temperature and wind strength or direction.

Given the extraordinary velocities that are reached during some windstorms, it is hardly surprising that they represent, along with earthquakes, the greatest harbinger of destruction on the planet. On the 18-point Beaufort Wind-force Scale—created by Irish hydrographer Francis Beaufort in 1805 and later expanded—gale-force winds register between points 7 and 10, corresponding to sustained wind speeds of between 52 and 102 kilometres an hour. In general terms, these are the sorts of wind speeds that might be expected in tropical storms or robust mid-latitude storms, such as periodically strike the UK and Europe. Point 11 on the scale, with wind speeds between 104 and 117 kilometres an hour, is defined as 'storm force', while all points above this relate to the progressively higher wind speeds encountered primarily in tropical cyclones. Point 18 on the scale involves sustained wind speeds in excess of 222 kilometres an hour, and is reserved for the most powerful tropical cyclones—major hurricanes in the Atlantic and super-typhoons in the Pacific. In 2015, Hurricane Patricia achieved an extraordinary one-minute, sustained wind speed of 345 kilometres an hour, which remains the world record.

While sustained wind speed is a good measure of the destructive capacity of a windstorm, this can be significantly magnified by peak gusts. Discounting the extraordinary speeds encountered in tornadoes, the most powerful gust ever was measured in 1996 at Barrow Island, off the north-west coast of Australia, during the passage of Cyclone Olivia, where the anemometer (wind speed recorder) registered a three-second gust of 408 kilometres an hour.

Windstorms can strike pretty much anywhere on the planet, and at any time. In different parts of the world, however, they tend to be more common during particular months or seasons. The Atlantic hurricane season, for example, officially runs from 1 June to 30 November, although this doesn't mean that hurricanes can't occur outside this window. If there is one thing we can be certain of, it's that nature does not play by our rules, which explains why hurricanes have formed as early as—or as late as—January. In the western Pacific, typhoons can form at any time of year. In the UK and Europe, powerful extra-tropical cyclones typically strike between October and March, but damaging storms can happen during any month.

The damage potential of wind can be astonishing, and communities caught in the most powerful tropical cyclones, or that find themselves in the path of a major tornado, can be all but obliterated. The threshold for notable wind damage begins at speeds of around 80 kilometres an hour. Above this, tiles can be torn from roofs, chimneys damaged, and branches broken off. At speeds in excess of 120 kilometres an hour, trees are toppled, mobile homes destroyed, and large chunks of debris hurled about. Wind speeds of more than 160 kilometres an hour cause serious damage to even well-constructed buildings, particularly to roofs; and many trees will be downed. Imagine, then, the colossal level of destruction caused by the most powerful tropical cyclone, wherein sustained wind speeds are in excess of 250 kilometres an hour, or by the most intense tornadoes, which can bring winds travelling at more than 500 kilometres an hour.

Damage and loss of life due to windstorms are not always a consequence of the wind itself, far from it. Extreme winds are often accompanied by snow or heavy rain. Tropical cyclones, in particular, can bring mammoth amounts of rain, leading to flooding on a colossal scale. In 1998, slow-moving Hurricane Mitch dumped up to 1.3 metres of rain in just a couple of days across Honduras and Nicaragua, leading to extensive flooding and thousands of landslides. More recently, in 2017, Hurricane Harvey—the wettest hurricane ever to make landfall in the United States—dropped 38 *trillion* gallons of rain over parts of Texas across a four-day period, leading to the flooding of hundreds of thousands of homes. Blizzards driven by extreme winds can also cause serious damage and disruption, as can windstorms that bring freezing rain, which coats everything with a thick layer of ice that can bring down power lines and make travel impossible.

In 2021, Winter Storm Uri—one of the worst winter storms to hit the USA in modern times—brought snow, ice, tornadoes, and temperatures below minus 30°C, which contributed to widespread power blackouts. In December 2022, Winter Storm Elliot left nearly 1.5 metres of snow in places, and more than seven million people in the US and Canada without power.

One of the biggest secondary effects of a windstorm can turn out to be the deadliest—the storm surge. Not only can a powerful storm cause waves up to 15 metres in height, but as it approaches a coastline it drives the water before it, pushing up the level of the sea. In addition, the low atmospheric pressure associated with the storm allows the sea to rise up beneath it, by 1 centimetre for every 1 millibar fall in pressure. Storm surges, especially those driven by intense tropical cyclones, can achieve heights of several metres, and the biggest in recent decades was that associated with Hurricane Katrina, which steered an 8.5 metre surge into the city of New Orleans in 2005. The impact of a major storm surge is barely distinguishable from that of a tsunami, and the damage and loss of life can be comparable. The storm surge brought by Cyclone Bhola to low-lying Bangladesh in 1970, resulted in at least 300,000, and possibly up to half a million, deaths, making it the deadliest windstorm on record.

Every year sees hundreds of windstorms, bringing damage and destruction anywhere from the equator to polar latitudes. In 2021, major windstorms in Europe and the tropics resulted in economic losses totalling US$94 billion, a substantial chunk of this—more than US$75 billion—due to damage and disruption caused by the impact of Hurricane Ida on Cuba and the US state of Louisiana. Once upon a time, such huge losses would have been unthinkable, but they are becoming increasingly commonplace, driven by greater concentrations of people and wealth, and by storms that global heating is making more powerful, wetter, and slower moving, so that they linger over one place for longer.

As with all geophysical catastrophes, the toll in generally better prepared developed nations tends to be a monetary one, while in Majority World countries, higher death tolls signal the severity of a storm. The costliest windstorm in history was the aforementioned Hurricane Katrina, which struck the US city of New Orleans in 2005, taking close to 2,000 lives and costing the national economy US$173 billion. The high death toll, in one of the world's richest nations, is a damning indictment of the high levels of poverty and enormous deprivation that coexist with extreme wealth here. Winter Storm Uri, introduced earlier, which battered much of the southern USA and northern Mexico during February 2021, is also worth a mention. Economic losses have been widely publicized as being of the order of US$25 billion, but some reports suggest that the storm could have resulted in damage and disruption totalling US$195 billion in Texas alone, taking it above Katrina in the costliest windstorm stakes. Cyclone Lothar, which rampaged across the UK and Europe in December 1999, remains the most damaging storm to strike the region in modern times, resulting in losses of US$15 billion.

Scourge of the tropics

The terminology around extreme windstorms that roam the tropics, between 5° and 30° north and south of the equator, can be confusing, so let's get this out of the way first. All organized systems of clouds and thunderstorms in the tropics, which spiral rapidly around a low-pressure centre of upwelling warm air, are known as *tropical cyclones*. In their infancy they are known as *tropical depressions*. These have maximum sustained wind speeds of less than 60 kilometres an hour, which are robust, but not powerful enough to cause notable damage. Once the sustained wind speed rises above this threshold, the depression officially becomes a *tropical storm*, which has the potential to be damaging. The critical sustained wind speed threshold, however, is 120 kilometres an hour, above which a storm can result in significant damage and loss of life. In the Atlantic and north-eastern Pacific such powerful storms are referred to as hurricanes; in the rest of the Pacific as typhoons; and in the Indian Ocean as cyclones. Just to add to the confusion, all powerful tropical cyclones are referred to as hurricanes in some quarters, especially in the more populist elements of the media.

Tropical cyclones come in all sizes, the biggest reaching up to heights of almost 10 kilometres and having widths in excess of 2,000 kilometres. Size isn't everything, however, and some of the most powerful and devastating storms have been small but very intense. For reasons associated with the Earth's rotation, tropical cyclones do not form within 5° of the equator. In addition, the Coriolis Effect ensures that north of this line they rotate counter-clockwise, while in the southern hemisphere they spiral in the opposite direction. Both north and south of the equator, tropical cyclones typically start off moving from east to west. In the northern hemisphere, they then curve around to the north-west, and even, ultimately, to the north-east. In the southern hemisphere, they tend to mirror this, curving first to the south-west and, in some cases, backing around to trend south-east. An important exception is the northern Indian Ocean where cyclones can follow a broadly northerly track into the Bay of Bengal or Arabian Sea.

Storms in the tropics require warm oceans to form and build, the energy that drives them coming from evaporation from the sea surface. To form at all, the sea surface temperature (SST) must be at least 25.5°C. Higher temperatures, in excess of 26–7°C will, however, encourage more rapid intensification. High SSTs along the track of a hurricane will also act to maintain intensification, leading, eventually, to a more powerful, and potentially more dangerous, storm. Once a tropical cyclone has formed, its growth will also depend critically on something called *vertical wind shear*, which is a change in either wind speed or direction, or both, with altitude. High vertical wind shear will chop up a potential cyclone before it really gets going, while low wind shear will cause minimal disruption to the developing storm.

The most powerful tropical cyclones—those with sustained winds above 120 kilometres an hour—are given names, and every year 40–50 of these wander across the world's three ocean basins. Despite hogging the news, only about 15 per cent occur as hurricanes in the Caribbean and North Atlantic, the majority happening in the Pacific and Indian oceans. A proportion of storms never make landfall, and although they may cause challenging sea conditions for shipping, they can live and die with few consequences for humankind. Those that do strike land, however, can exact enormous tolls in terms of destruction and loss of life and livelihood. The damage potential of a tropical cyclone heading for landfall can be gauged, to a large degree, by the expected wind speeds. Atlantic hurricanes are classified by wind speed using the *Saffir-Simpson Hurricane Wind Scale* (SSHWS) (Table 4), which has five categories. The lowest of these, Category 1—or simply Cat1—is allocated to the weakest storms, having sustained wind speeds of between 119 and 153 kilometres an hour. Sustained wind speeds increase progressively for higher categories, up to Cat5, for which the one minute maximum sustained wind speed reaches 252 kilometres an hour or more.

Cat3 to Cat5 storms are known as major or severe hurricanes, and it is these, all other things being equal, that cause the most damage and loss of life. A Cat5 storm, for example, is violent enough not only to uproot or snap all trees, but even to tear the bark off some. Any building not well constructed from solid concrete or with a steel frame will be destroyed or severely damaged. The terrible impact of a Cat5 storm on an urban centre is evidenced by the aftermath of Hurricane Andrew, which struck just south of downtown Miami in 1992. In total, more than 63,000 homes were destroyed and 101,000 damaged, leaving 175,000 people homeless (Figure 12). Different scales are used to

Table 4. The Saffir-Simpson Hurricane Wind Scale

Category	Max one minute sustained wind speed (km/h)	Damage and likely consequences
1	119–53	Very dangerous winds resulting in some damage. Well-constructed frame buildings could see damage to roofs, gutters, and plastic cladding. Large branches of trees snapped and shallow-rooted trees may be toppled. Extensive damage to power lines and poles likely to result in power outages that could last several days.
2	154–77	Extremely dangerous winds resulting in extensive damage. Well-constructed frame buildings could sustain major roof and side damage. Many shallow-rooted trees snapped or uprooted, blocking numerous roads. Near total power loss expected, with outages lasting from several days to weeks.
3	178–208	Devastating damage. Major damage to well-constructed frame buildings, including roof and gable end removal. Many trees snapped or uprooted, blocking numerous roads. Water and electricity supply unlikely to be available for several days to weeks.
4	209–51	Catastrophic levels of damage. Severe damage to well-constructed frame buildings, including loss of complete roof structures and/or exterior walls. Most trees snapped and power-line poles downed. Power outages expected to last weeks to months. Most of affected area uninhabitable for weeks to months.

| 5 | >252 | Catastrophic levels of damage. A high proportion of frame buildings destroyed due to roof loss and wall collapse. Fallen trees and power-line poles cut off residential areas. Most of affected area uninhabitable for months. |

Notes: The SSHWS classifies hurricanes according to their sustained maximum wind speeds. Here, the five categories are shown alongside expected damage and likely consequences.

Storm force

12. Hurricane Andrew (1992) was one of only four hurricanes to make landfall on continental USA at Cat5 strength. The storm destroyed or damaged more than 160,000 homes in the Miami area, and left 175,000 people homeless.

categorize tropical cyclones outside the Atlantic and north-eastern Pacific, but all are similarly based on sustained wind speed. In the western Pacific, a storm with sustained wind speeds of 222 kilometres or more is known as a *super typhoon*, while in the North Indian Ocean, sustained wind speeds of 222 kilometres or more define a *super cyclonic storm*.

As well as coming in all sizes, tropical cyclones also display a wide range of characteristics. At 2,200 kilometres across, and with a world record low pressure of 870 millibars at its core, Super-typhoon Tip—which battered Japan in 1979—was the biggest, and most intense, tropical cyclone on record (Figure 13). In contrast, with storm-force winds extending out to less than 10 kilometres from the centre, Tropical Storm Marco (Caribbean, 2008) was the smallest. The wettest storm was Cyclone Hyacinthe, which dumped 6 metres of rain on the Indian Ocean island of Réunion over the course of 15 days. One of the most worrying characteristics is how rapidly a storm can wind itself up into something really savage. So-called *rapid intensification* is defined as an increase in the sustained wind speed of at least 55 kilometres an hour in a 24-hour period. In reality, it can involve a far bigger hike. In 2015, the sustained wind speed of Hurricane Patricia exploded from 135 kilometres an hour (Cat1) to 330 kilometres an hour (Cat5) in less than a day, making it the most powerful tropical cyclone on record. Such an extraordinarily rapid

13. At 2,200 kilometres across and with a central low pressure of just 870 millibars, Super-typhoon Tip (1979) holds the record for the biggest and most intense tropical cyclone on record. Here it can be seen in the centre of the image, with Typhoon Sarah to the left.

strengthening can cause serious problems, both for forecasters and for those planning and enabling evacuation.

As powerful storms exit the tropics and head polewards over cooler seas, they typically lose much of their power. Nonetheless, they can remain hugely destructive. In 2012, Hurricane Sandy caused widespread damage to New Jersey and New York; and in October 2022, Hurricane Fiona brought Cat2 strength winds to Nova Scotia, making it the strongest storm ever recorded in Canada.

The storms in the middle

Strictly speaking, storm Fiona was no longer a hurricane when it battered eastern Canada. By this stage it had transitioned to an *extra-tropical cyclone*, not that this distinction meant much to the many thousands of people affected. This transitioning process is quite common, allowing former tropical cyclones to continue as powerful storms at higher latitudes. In the Atlantic, it is not unusual for them to cross over to the UK and Europe, where they can still be damaging. This is only one way, however, that extra-tropical cyclones—also known as mid-latitude storms—can form. More typically, hundreds of these large-scale storm systems develop every year, solely at mid-latitudes, typically between 30° and 60°. They originate as waves along weather fronts, which develop into low-pressure systems that progressively intensify from weak depressions into powerful windstorms as they travel (generally) from west to east. This process, known as *cyclogenesis*, can happen extremely rapidly, and the term *explosive cyclogenesis* is used to describe a storm whose central pressure is falling at a rate of more than 1 millibar an hour. Then the storm becomes a so-called 'weather bomb', which you might have heard referred to in weather forecasts or read about in one of the more excitable newspapers. Such storms can have central low pressures comparable to those of tropical cyclones, and in 1986 a storm system close to Iceland registered a core pressure of 920 millibars, the same as found at the heart of a Cat5 hurricane.

Unlike tropical cyclones, for which warm ocean surface waters provide the energy source, mid-latitude storms get their energy from large horizontal differences in atmospheric temperature—for example where frigid polar air comes up against warmer humid air from nearer the tropics. Mid-latitude storms can be as wide as the biggest tropical cyclones and cause damage or destruction across half a continent. Although maximum wind speeds are nowhere near as high, hurricane-force winds are not uncommon. Analysis of records of the great storm that struck the UK in 1703 suggest that it was comparable in strength to a Cat2 hurricane, bringing gusts that may have exceeded 270 kilometres an hour. In 1962, Typhoon Freda transitioned into the mid-latitude Columbus Day storm, bringing Cat3 winds that caused heavy rain and widespread damage across the US states of Washington and Oregon.

As is also the case for tropical cyclones, however, the damaging impact of mid-latitude storms is not limited to the strength of the wind. In the USA, winter storms often bring widespread blizzard conditions, or freezing rain. In 2021, blackouts caused by Winter Storm Uri affected close to 10 million people, and resulted in almost 300 deaths. In July 2016, a not especially powerful storm brought sustained torrential rains that caused devastating flash flooding across northern China, which affected more than 15 million people and destroyed 130,000 homes.

In those parts of the world where mid-latitude storms are common, they have taken on local names. In the north-west of the United States, the powerful storms that build in intensity as they track eastwards across the North Pacific are known as *Big Blows*. In the north-east of the country, similarly powerful storms that bring heavy rain, blizzards, and bitter north-easterly winds are termed *Nor'easters* or *East Coast Lows*. In the southern hemisphere, powerful storms that can develop over Argentina and Uruguay are known as *Sudestadas* (South-east Blows). So-called *Maritime Lows* or *East Coast Lows* affect the south-east coast of

Australia, bringing storm-force winds and extreme rainfall and floods, usually during autumn and early winter. In the first half of 2022, such a storm brought some of the most devastating flooding in the country's history. Worst hit were coastal New South Wales and southern Queensland, where Brisbane received 40 centimetres of rain over a three-day period.

Mid-latitude storms are also a common feature of the October to March weather across Europe, where they are rather unimaginatively referred to as, simply, European windstorms. In almost every case, the storms track west to east across the North Atlantic, and some may even begin as Nor'easters on the other side of the 'pond'. Strengthening as they go, they normally track across northern Scotland or the Norwegian Sea, where they can do little harm. On occasion, however, they follow a more southerly track, which means they can impact severely on major population centres in Ireland, the UK, France, the Low Countries, and Germany. In a typical year, such storms cause damage amounting to a couple of billion euros. They also cause the bulk of insurance losses, hardly surprising when you consider that every year around 200,000 homes are damaged by wind in the UK alone.

Notable European windstorms include the Great Storm of 1987 (Figure 14), which brought sustained Cat1 hurricane-force winds in excess of 121 kilometres an hour to the UK, and felled an estimated 15 million trees. Perhaps even more powerful was Cyclone Daria (also known as the Burns' Day Storm), which in January 1990 pummelled the UK, and later the Low Countries and Germany, with sustained winds of almost 140 kilometres an hour. One of the most dangerous aspects of European windstorms is their tendency to cluster, so that two or three may arrive within a few days of one another. Over Christmas 1999, for example, Cyclone Lothar—widely regarded as the worst European windstorm of the 20th century—was followed, less than 36 hours later, by Cyclone Martin. Lothar is the most damaging European windstorm on record, costing economies more than 15 billion euros. Martin,

14. **The Great Storm of October 1987 brought hurricane-force winds to southern parts of the UK, and felled around 15 million trees.**

which followed a slightly more southerly track, added another 6 billion euros to this. In a similar manner, Cyclone Kyrill—another devastating storm with gusts of up to 250 kilometres an hour—arrived in January 2007, just four days after Cyclone Per.

Going back to well before storms were named, an anonymous European windstorm in January 1953 was by far the deadliest of modern times, and it wasn't just the speed of the wind that caused problems. The worst storm to strike the northern UK for centuries sank at least 10 steamers and trawlers, together with a ferry crossing the Irish Sea, taking more than 230 lives in all. When it had finished with the shipping, the storm moved southwards into the North Sea. Here, a combination of wind and the deep low pressure at the storm's centre caused a huge storm surge to develop, amplified by a high spring tide. As the storm centre continued to move southward, the record surge—more than 3 metres in height—spilled over onto adjacent low-lying coastal areas. Flood defences failed in many places along the east coast of

England, inundating thousands of properties and drowning more than 300 people. The situation was far worse, however, across the North Sea, where the breaching of protective dykes allowed the sea to pour across more than 1,000 square kilometres of the Netherlands, much of it below sea level. By the time the sea retreated, nearly 50,000 buildings had been flooded and more than 1,800 people were dead.

Tornadoes and their ilk

While tropical cyclones and mid-latitude storms can be thousands of kilometres across and persist for several days, other expressions of extreme wind are more localized and shorter lasting. They can, nonetheless, be similarly deadly and destructive, albeit on a smaller scale. Right at the top of the list, in terms of their calamitous potential, come the violently rotating tubes of wind known as tornadoes, or—colloquially—twisters. These can form almost anywhere on the planet, and tornadoes have taken lives in countries as far apart as Australia, South Africa, Argentina, Bangladesh, China, Japan, Russia, and Italy. Their main hunting ground, however, is undoubtedly North America, and especially the central and south-eastern states of the USA, a broad region termed 'tornado alley', where they cause loss of life every year.

In 2021, almost 400 tornadoes were reported across the United States, together taking 104 lives, making it the deadliest tornado season for a decade. As elsewhere on the planet, most US tornadoes are relatively benign, having wind speeds of less than 180 kilometres an hour, and diameters, where they come into contact with the ground, of less than 80 metres. Typically, such storms travel just a few miles before dissipating, but they can still cause significant damage if they encounter populated centres during their short existence. Most loss of life comes from far more powerful versions that can be 3 kilometres across, have wind speeds approaching 500 kilometres an hour, and can remain in contact with the surface for distances of more than 50 kilometres.

Such monsters can wreak havoc in urbanized areas, leaving behind long trails of total destruction. Most US tornadoes form between March and June, when warm, humid air from the tropics battles with cold polar air; but they can and do happen in every month of the year. Tornadoes can also form, from landfalling hurricanes, between June and November.

The most commonly used scale for classifying tornadoes is the *Enhanced Fujita Scale*, which ranges from EF0 up to EF5. Tornadoes that score a zero have winds that are sufficient to cause damage to trees, but not much else. At the top of the scale, an EF5 is capable of ripping buildings wholesale from their foundations and even damaging steel and concrete highrises.

While tornadoes do occur singly, it is common for them to form in clusters. This is because, if conditions are right for the formation of one tornado then they are equally suited to the formation of many. Such clusters, known as 'outbreaks' in the US, are responsible for the highest number of deaths, a 'super' outbreak of 148 tornadoes in 1965 taking more than 300 lives. This isn't, however, to demean the potential impact of single tornadoes, which can prove to be at least as devastating. In 1989, a single, huge EF4 twister, 1.5 kilometres across, travelled for 80 kilometres through the Dhaka region of Bangladesh, killing 1,300 people and leaving 80,000 homeless. In 1925, the so-called 'Tri-State' tornado (estimated to have been an EF5)—the deadliest in US history—left a 350-kilometre trail of destruction across Missouri, Indiana, and Illinois, taking close to 700 lives and obliterating 15,000 homes.

Most tornadoes form during thunderstorms, the most powerful—EF3 to EF5—developing from giant *supercell* storms that have a rotating core known as a *mesocyclone*, which typically extends from around 2.5 kilometres above the surface to heights of almost 10 kilometres. A tornado develops when heavy rainfall pulls with it downdraughts of dense, cool, air, which focuses the rotating mesocyclone and drags it down towards the surface.

This results in the formation of rapidly rotating *funnel clouds* that extend below the storm's cloud-base, and that are visible because the reduced pressure at the heart of the vortex causes water vapour to condense. If and when the funnel cloud touches the ground, it becomes a tornado, pulling up dust, soil, and debris that further increases its visibility. A tornado ceases to be worthy of the name once the cold downdraughts cut off the storm's supply of converging warm air, which normally happens after 10–20 minutes. During the 'Quad-State' tornado outbreak of December 2021, however, which affected Tennessee, Kentucky, Missouri, and Arkansas, and destroyed the Kentucky town of Mayfield, one tornado may have persisted for more than two hours as it tracked across Tennessee and Kentucky.

Although tornadoes are the best known, and certainly most photogenic, localized windstorms, there are others. *Derechos*, which are the straight-line equivalents, are also associated with severe thunderstorms, and can bring hurricane-force winds, damaging hail, and torrential rain capable of causing localized flash flooding. They are the extreme examples of 'squalls', bands of storms that result in the sudden arrival of strong winds and rain, and which can play havoc with small boats. The most intense Derechos are characterized by gusts in excess of 90 kilometres an hour, sustained for several hours, which are damaging across thousands of square kilometres of countryside. Derechos normally form in the summer months, associated with convection-related thunderstorms, and, like tornadoes, are most common in North America, where they can have severe impacts on power infrastructure. Like tornadoes, they can happen anywhere on the planet, and a Derecho in 2002 killed eight people and injured 39 others in the German capital, Berlin.

Other potentially damaging winds include those confined by geography to specific parts of the planet. The best known is probably the *Mistral*, the powerful, cold wind that periodically blows into the south of France from the north-west, and which commonly reaches sustained speeds of more than 65 kilometres an hour. There are dozens of other potentially damaging winds arising from local or regional circumstances of meteorology and geography, including the *Sirocco*, which originates in the Sahara Desert, and California's dry *Santa Ana* winds, which can make more problematical attempts to bring the state's increasingly dangerous wildfires under control. In desert regions, cyclones or thunderstorms can initiate damaging sandstorms, such as Sudan's *Haboob* or the Sahara's *Simoom*.

Tempests in a hotter world

With added heat to further energize the weather machine, you might expect a hotter planet to be a windier one, but the true picture is more nuanced. One study has revealed that the world has indeed become windier since 2010, by a not insignificant 7 per cent. In the preceding couple of decades, however, average wind speeds across the planet actually fell by a couple of percentage points, so it is not obvious if the recent rise is linked to global heating.

As far as tropical cyclones are concerned, however, the trends are clearer and the news is bad. Average wind speeds are climbing by 3–4 per cent for every 1°C rise in global average temperature, but, more importantly, so are the peak wind velocities responsible for most damage. Even worse, the probability of a hurricane becoming a major Cat3 storm has been rising by around 8 per cent a decade, and will continue to do so. This is revealed by an already apparent increase in the number of more powerful storms in all ocean basins. Wind speeds are increasing so fast that there is even talk of modifying the SSHWS to include a new Cat6, into which the most powerful Cat5 storms would be placed. These might include the 18 storms since 2010 that have seen wind speeds touch an extraordinary 286 kilometres an hour.

In the Atlantic, a recent study has revealed that the most active hurricane seasons are now twice as common as they were in the 1980s, raising the potential for increased destruction and loss of life. Tropical cyclones are getting much wetter too, dramatically increasing their ability to cause major flooding. New evidence also suggests that higher SSTs due to global heating may be driving an increase in the number of storms that undergo rapid intensification, which has serious disaster preparedness ramifications.

And there is more grim news. When a tropical cyclone makes landfall it usually dissipates rapidly, but it seems that due to global heating storms are now taking twice as long to decay as they did 50 years ago, carrying the potential for serious damage and loss of life much further inland. On top of this, some tropical cyclones are moving more slowly, with a tendency to stall, meaning that the length of time a particular location may be subjected to extreme winds or torrential precipitation is also increased. As global heating causes the tropics to expand, the warm seas required for the formation and sustenance of tropical cyclones are being encountered further north and south. Storm tracks are, as a consequence, shifting polewards so that they reach peak activity at higher latitudes, placing more, often less prepared, coastal cities under greater threat.

Broad patterns of tropical cyclone activity may also be modified if there are significant changes in the prevalence of warm-phase (El Niño) and cold-phase (La Niña) conditions in the tropical Pacific Ocean, which tip back and forth like a see-saw. When El Niño conditions prevail, Atlantic hurricane activity tends to be damped down, while tropical cyclone activity is increased in parts of the Pacific. In contrast, when La Niña is dominant, the opposite is observed. Any changes in the frequency of El Niño versus La Niña, as global heating progresses, is therefore likely to be reflected in the relative occurrence of active tropical cyclone seasons in the two ocean basins.

The tracks of mid-latitude storms are also forecast to move closer to the poles as our world continues to heat up. While storm numbers are not expected to rise, and may even fall a little, the number of extreme storms in the winter months is projected to increase. Across the UK and Europe, warmer Atlantic surface waters at higher latitudes may increase the intensity of autumn storms that have transitioned from hurricanes. According to the Association of British Insurers, a 1.5°C global average temperature rise will result in UK windstorm insured losses increasing by 11 per cent, this figure climbing to 23 per cent for a 3°C rise. Surges associated with both tropical cyclones and mid-latitude storms will become progressively more damaging as sea level continues to rise, increasing the coastal flood threat and pushing it further inland.

Chapter 5
Fire and flood

The rise and rise of the wildfire

When a wildfire guts half a village on the edge of London, you know something unprecedented is happening to our world (Figure 15). In years past, wildfires seemed solely the preserve of Australia and California, but suddenly they are everywhere, even—astonishingly—Greenland. Along with every species of extreme weather, their occurrence and impact has exploded in the past five years or so, bringing devastation and loss of life to Canada, western USA, Russia, Spain, France, southern Europe, and elsewhere.

The causes of wildfires remain the same as always—lightning, the Sun's rays refracted through bottles or broken glass, barbecues, discarded cigarettes, even intentional ignition. But the huge increase in prevalence is directly related to the longer and more intense heatwaves driven by global heating, and the tinder-dry conditions they foster. Wildfires used to be confined largely to areas of low population. More recently, they have become a problem at what is called the wildland–urban interface, where countryside abuts against larger settlements. And today, fires have come to seriously threaten major conurbations, including Los Angeles and Sydney, and have wiped entire communities off the map. Inevitably, as wildfires intrude into more populated areas,

15. Destructive wildfires are beginning to happen in countries where they were previously all but unknown. Image shows properties destroyed by a wildfire on the outskirts of London, UK.

more lives will be lost. Fires stoked by the scorching heat of summer 2018, and spurred on by gale force winds, took more than 100 lives in the Attica region of Greece, some burned alive in their cars and homes. In Australia, the 2019/20 wildfire season saw more than 6,000 properties destroyed and left 34 people dead.

Although 2022 was bad for wildfires, the previous year was the worst for which a full record is available. It began in earnest with the development, in late June 2021, of a colossal heat 'dome' across much of western North America, which saw temperatures exceed an unprecedented 49°C in British Columbia and the obliteration by fire of the Canadian town of Lytton. In the months that followed, wildfires continued to rage across western Canada, and all down the US west coast as far as the Mexican border. In California alone, more than 8,000 fires burned an area in excess of 10,000 square kilometres—about equal to the size of Jamaica. Included in this figure was the historical gold rush town of Greenville, completely erased by the Dixie fire, the largest in the state's history, which took three months to bring under control. By the year's end, almost 50,000 fires had reduced more than 26,000 square kilometres to ashes, an area almost equal to the size of the state of Massachusetts. Even in the final month of the year, wildfires were still happening in the US, Colorado experiencing huge blazes that destroyed 900 buildings and forced thousands from their homes.

At the same time that North America was burning, so was Russia, where temperatures within the Arctic Circle climbed above 30°C in June and July 2021, spawning huge wildfires across the great coniferous forests of the *taiga*. By August, the Siberian fires were the largest on the planet, their smoke reaching the North Pole. Wildfires hit Europe hard in 2021 too, although 2022 proved to be even worse. Sustained drought conditions, and summer temperatures of more than 45°C, resulted in fires burning right across the southern part of the continent, from Portugal to Greece. By the time the flames had been damped down, the blazes had, together, burned an area almost four times the size of Greater London—a new record for Europe. The first half of 2023 saw unprecedented fires across Canada that, by June, had burned eleven times the long-term average for the time of year—an area approaching the size of Austria.

From being a minor nuisance in all but a few parts of the world, wildfires now constitute a deadly and destructive hazard that can make a sudden appearance almost anywhere on the planet. Taking just a few seconds to get going, they can take many months to bring under control, and can even smoulder unnoticed over the winter months to re-emerge in the spring. The most powerful can generate their own weather too, becoming self-sustaining by building giant *pyrocumulonimbus* clouds overhead, which send down lightning strikes that start new blazes. Wildfire weather can also include unstoppable fire tornadoes, which rampage across the landscape bringing wind speeds of 200 kilometres an hour.

As wildfires get ever larger and more persistent, and the smoke they generate becomes so voluminous that it can cross entire continents, they bring growing health problems. In spring and early summer 2023, smoke from Canadian wildfires shrouded New York, Washington, DC, and Philadelphia, halting flights, closing schools, and resulting in the worst air quality on the planet. Every year, more than 30,000 people lose their lives as a consequence of wildfire pollution, with firefighters particularly vulnerable. An even bigger issue is the vast quantity of carbon released from the wholesale burning of bush and forest. In 2021, a colossal 6.5 billion tonnes of carbon dioxide was added to the atmosphere by wildfires—close to 150 per cent higher than the fossil fuel emissions of the entire EU in 2020. This, of course, is a positive feedback effect. The hotter and drier global heating makes our planet, the more fires there are. The more fires there are, the more carbon dioxide is released, which—in turn—makes the world even hotter. Added to this, major wildfires also reduce the density and size of the trees that eventually grow back, so that they do not absorb as much carbon as those destroyed.

This is not, however, entirely inevitable. Many wildfires do not start naturally but are a consequence of deliberate human action. Burning virgin, mature, forest in order to make space for ranching or agriculture remains routine in a number of countries, notably Brazil. Across the planet, an area of primary tropical rainforest equivalent to a football pitch was lost, much of it by burning, for every *second* of 2019. This added 1.8 billion tonnes of carbon dioxide to the atmosphere—equal to the annual emissions of half of the world's cars. In Brazil alone, an area of the Amazon Rainforest larger than Cyprus was lost in the year to July 2021, much of it as a result of intentional burning. With a new administration in charge, and with an international agreement in place to stop deforestation by 2030, there is some hope that emissions from deliberate burning will begin to fall soon—and not before time. In the meantime, measures can be taken to try to keep the lid on natural wildfires. Controlling them once they get going is becoming increasingly difficult, so efforts need to be focused on prevention. Measures that can help include intentional, cool-season burns, aimed at destroying as much as possible of the old wood, brush, and leaves that provide the raw material for hot season fires.

Water world

To anyone caught up in one, a flood is a flood, although they do, in fact, take many forms. The broadest distinction is between inland floods and those that affect coastlines and coastal communities. Inland floods, in turn, can be subdivided into those caused by overtopping rivers and those that arise from water flowing across the ground surface. At the coast, flooding can occur due to exceptionally high tides, as a consequence of storm surges, or, increasingly, because of permanent inundation caused by rising sea levels.

Historically, floods have been one of the most devastating of all geophysical hazards, in terms both of impact and death toll. Topping the list are the 1931 Yangtze-Huai river floods, which affected more than 50 million people and directly took at least 150,000 lives. Hundreds of thousands more lives were lost as a result of rampant cholera and famine following the wholesale destruction of crops. Some accounts propose that the death toll could have been as high as four million, which would make the event the biggest geophysical catastrophe of all time, but this figure cannot be validated. Half a century or so earlier, in 1887, the Yellow River burst through its retaining dykes, flooding an estimated 130,000 square kilometres of adjacent land—an area about the size of England. The resulting death toll, including those killed by disease and starvation, is often quoted as being anywhere from 900,000 to three million but, once again, it is not possible to narrow down or corroborate these figures.

Although flooding on such a prodigious scale might seem hard to believe, we have more recent evidence to show that this is not the case. The 2022 Pakistan floods, for example, swamped between 10 and 12 per cent of the country—at 85,000 square kilometres, an area twice the size of Switzerland. While the death toll was restricted to less than 2,000, the number of people affected is estimated at more than 33 million, of which more than two million were left homeless. The material cost of the flooding included more than 2.3 million homes and 22,000 schools destroyed or damaged, and more than a million head of livestock lost. The consequences for Pakistan's economy were colossal, estimates of the cost of the event and reconstruction adding up to US$30 billion, a full 10 per cent of the country's gross domestic product (GDP).

While exceptional in terms of scale and impact, what happened in Pakistan in 2022 constitutes just one of many disastrous flood events that have plagued every part of the planet in recent years. During late 2019 and early 2020, some of the heaviest rainfall on record brought severe flooding to parts of the UK. In the summer of 2020, monsoon flooding on an immense scale affected close to 60 million people in India due to the Brahmaputra river bursting its banks. A further 50 million were impacted by widespread flooding in southern China, and getting on for a million more as extreme rains triggered massive flooding and landslides in east Africa. Spring 2021 saw parts of eastern Australia swamped by floodwaters, while major flooding continued to affect many parts of China. In July of the same year, Europeans were shocked by the violence and destructive power of the flash floods that roared through parts of Germany and Belgium (Figure 16), tearing communities apart and taking 243 lives. The picture didn't improve in 2022 either, south-east Australia being hit again by some of its worst ever flooding, and heavy rains and a storm surge bringing severe flooding across south-west Florida.

The aforementioned instances represent just a small fraction of the serious flood events that have proved in recent years to be the

16. Damage to the Belgian town of Pepinster, caused by severe flash flooding in July 2021.

bane of communities, both large and small, in every corner of the world. Since 1980, it is estimated that floods have cost more than US$1 trillion, while a recent study puts the figure for economic losses due to flooding between 2000 and 2015 at US$651 billion. If it seems that there are more floods today than there used to be, that's because it's true. This shouldn't really come as a surprise as a hotter world is a wetter one too. As the atmosphere heats up, so it contains more water in the form of vapour—up 7 per cent since pre-industrial times—feeding increased rainfall across land areas, and coming in more intense bursts than it used to. A study published in 2018 revealed that, globally, incidents of extreme rainfall and flooding had increased four times since 1980, and 50 per cent since 2010. Between 2000 and 2018, almost 300 million people were affected by more than 900 flood events, and this figure is set to climb far higher as global heating continues.

In the US alone, increases in the volume and intensity of rainfall raised the cost of flood damage between 1988 and 2017 by US$73 billion. In the UK, around 2.4 million homes are at risk from river or coastal flooding every year, and a further 2.8 million susceptible to surface flooding. Something like one in six homes are located in areas where there is significant risk of flooding, an especially sobering statistic given that flood risk is only going to increase. So bad has the situation become in recent years that the chance of being flooded out of your home is now higher than the likelihood of your house being burgled.

For every 1°C temperature rise, global mean precipitation climbs by between 1 and 3 per cent. It is not entirely surprising, then, that analysis of notable flood events reveals that climate change has had a hand in many. In fact, attribution studies have demonstrated that 56 per cent of recent major floods or incidences of extreme rainfall have been made either more severe or more likely due to global heating. It would be wrong to suggest, however, that climate breakdown is the only cause of the inflating flood bubble. Land-use changes, notably the large-scale destruction of forests and the degradation of farmland, also play a significant role through increasing run-off into watercourses and along the surface. In many parts of the world the impacts of floods are being made worse by the growth of urban centres sited in flood-prone areas. In particular, the rapid expansion of mega-cities, especially in Majority World countries, often involves *ad hoc* building of poorly constructed properties in flood plains or on steep slopes easily destabilized by heavy rain and surface run-off. In developed nations, too, the lesson that building on flood plains inevitably ends badly seems not yet to have sunk in.

Rivers down below and rivers in the sky

A river floods when the amount of water entering it exceeds its capacity to shift it downstream. When this happens, levels rise until the containing banks are breached in one or more places, allowing excess water to spread out into the surrounding countryside. In an entirely natural situation, a river, at least in its more mature stages, is bounded by flat land built from the sediment deposited by previous floods. This so-called flood plain does exactly what it says, providing an area that holds excess water when river banks are breached. Unfortunately, in the modern world, the tendency has been to ignore or conveniently 'forget' what flood plains are there for. Instead of being left in their natural state to do their job, many in developed nations are buried under concrete and tarmac, given over to housing or industrial estates, or are even made home to critical infrastructure.

To free up flood plains for building, so-called 'hard' engineering measures are commonly inflicted on rivers, including dredging, 'channelizing' them within concrete walls, straightening out their courses, and building raised levees in an attempt to stop breaches of the banks. All of these actually make the situation worse, typically speeding up the flow, but often also restricting the volume of water that can be carried downstream in a given time-frame, and so facilitating flooding. 'Protective' levees, in particular, store up real trouble, because as the river bed silts up, they have to be built ever higher. Ultimately, this leads to the river flowing at a level elevated above the surrounding land, so that any breach has the potential to generate a sudden and potentially catastrophic flood. This is exactly what happened in 2005 when levees holding in the Mississippi failed under the onslaught of rains brought by Hurricane Katrina, swamping much of New Orleans, and also along the course of the Brahmaputra in Assam (India) during the 2020 monsoon season.

Despite, or because of, human intervention, river flooding is becoming more, not less, common. As touched on earlier, this is, to a large degree, driven by more intense rainfall. Even without the influence of global heating, the intensity of rainfall can be quite astounding. In 1956, more than 3 centimetres of rain fell in just 60 *seconds* in Unionville, Maryland (United States)—a world record. On the French island of Réunion, in the Indian Ocean, a passing cyclone dropped nearly 2 *metres* of rain during a single 24-hour period in January 1966.

Probably the main consequence of more short-lived, but extreme, cloudbursts, often associated with major thunderstorms, is an increase in the occurrence of flash floods. These can transform gently flowing rivers into raging torrents within minutes, leaving little time for warning, and are almost invariably lethal as well as hugely destructive. The ferocious German and Belgian floods of 2021 were only one example of several that year. August saw more than 80 people killed and almost 500 homes destroyed as powerful thunderstorms dumped as much as 1.2 metres of rain in 48 hours across Turkey's Black Sea region. In the same month, more than 50 centimetres of rain in 24 hours powered flash floods in parts of the US state of Tennessee, taking more than 20 lives.

Intense downpours are also responsible for surface flooding, whereby water overwhelms natural and artificial drainage systems and flows across the surface, with little permeating below. This type of flooding is common where soils have become saturated due to previous heavy rains, but also following droughts or extremely dry periods when it is difficult for rainwater to penetrate the baked surface layer. Surface flooding is becoming a growing problem in cities, where an almost continuous cover of concrete and tarmac acts as a barrier to infiltration, and drainage and sewerage systems often can't cope. The deadly consequences were played out in central New York City in September 2021, when torrential rain from the remnant of Hurricane Ida spawned severe surface floods that drowned people in their basement apartments.

Extreme rainfall can also trigger mudflows and landslides that can be extremely destructive and deadly. In 1998, for example, the torrential rains associated with Hurricane Mitch—the deadliest storm to strike Central America—triggered a massive mudslide on the flanks of Nicaragua's Casita volcano, which buried several villages along with more than 2,000 of their inhabitants. More recently, in early 2022, copious rains triggered landslides and mudflows on steep slopes, which killed 230 people in the poorer, marginal areas of the Brazilian city of Petrópolis.

Slower onset, but more sustained, flooding is a consequence of persistent heavy rain across a particular area, which—over time—overwhelms catchments, so that the rivers draining them cannot carry the extra water. When this happens, banks or levees are overtopped and flooding occurs that can, as in Pakistan in 2022, be extraordinarily extensive. It can be long-lasting too, so that floodwaters affecting New Orleans following the arrival of Hurricane Katrina at the end of August 2005 were still not completely gone by early October.

The sort of long-lasting rain that brings with it the threat of major flooding is often associated with the monsoons—the annual rainy seasons common to South and South-East Asia, Australia, West Africa, and some other parts of the world. Elsewhere, it can be brought by a variety of local or regional weather patterns that, for one reason or another, act to maintain heavy rainfall over several days. In some instances, perfect conditions for seemingly unending precipitation are the result of what are known as *atmospheric rivers* or 'rivers in the sky'. These long, narrow conveyor belts of air carry around 90 per cent of the moisture from the tropics to higher latitudes, adding up to a greater flux of water than that of the Amazon River. Atmospheric rivers are especially persistent, so that the rain over a particular region will just keep on falling, for days at a time, saturating the ground and bringing widespread flooding. Around a dozen such 'rivers' are active at any one time, so someone, somewhere, is always on the receiving end. In the UK, it was the Lake District that suffered severe flooding as a result, both in 2009 and 2015. In late 2021, the Canadian province of British Columbia was in the firing line, causing floods that effectively cut off the city of Vancouver. And in December 2022, San Francisco saw its wettest day for at least 150 years as an atmospheric river surged overhead.

Going under

Coastal flooding, like that occurring inland, has always been with us, driven by high tides or storm surges, and sometimes a combination of the two. Now, however, the threat is being augmented by global heating. I tend to think of the coast as marking the frontline of climate breakdown. Here, it is possible to measure directly the accelerating rise in global sea level, which bears a direct relationship to the thermal expansion of warming oceans and increasing melting of glaciers and polar ice sheets. These, in turn, are proxies for the progressive rise in the global average temperature caused by greenhouse gas emissions.

As sea level rises, so exposure to coastal flooding increases too, along with the threat of permanent inundation of low-lying topography. Over the course of the 20th century, sea level rose by around 20 centimetres at a rate, between 1900 and 1990, of 1.4 millimetres a year. In the last 30 years, however, this rate has increased dramatically to half a centimetre annually, and it continues to accelerate. Bearing in mind that just a 1-centimetre rise means that an additional six million people come under threat from coastal flooding, this is not good news.

It should be obvious that ever climbing sea levels progressively increase the risk of coastal flooding, as well as bringing more people under threat. High tides and storm surges will be that much higher, the frequency of flood events will shorten, and existing defences will become less effective. Coastal flooding can occur simply as a consequence of the level of the sea rising above the elevation of the land. In low-lying places like southern Florida, for example, the highest spring tides of the year—sometimes called King Tides—are causing problems as sea levels continue to climb. In 2017 and 2019, especially high tides flooded roads and gardens, causing damage and travel problems. At the moment, it is simply nuisance flooding, but it won't stay that way. King Tides provide a glimpse of the future, and ultimately, as the sea-level rise accelerates, the floods will come to stay.

The most destructive and lethal coastal flooding arises as a consequence of the overtopping or breaching of natural or engineered defences. Worldwide around 150 million people living on the coast are currently exposed to flooding, sometimes with appalling consequences. The 10-metre storm surge ahead of Cyclone Bhola, which overwhelmed southernmost Bangladesh in 1970, taking at least 300,000 lives, is a brutal reminder of the potentially colossal scale of flooding where defences are minimal or non-existent. Similarly, the 1953 North Sea floods provide testament to the destructive and lethal power of the sea in places where defences fail to protect populated land that is below, at, or barely above sea level.

In truth, no coastal city is immune to flooding by the sea, a fact forcefully demonstrated by the impact, in 2012, of Hurricane Sandy on New York. Here, a combination of record high tides and a storm surge in excess of 4 metres in places (Figure 17) resulted in the highest water levels ever recorded in the New York Metropolitan Area. Around 17 per cent of New York City was flooded—including hundreds of thousands of homes and more than 20,000 businesses.

Probably the best demonstration of what happens when coastal flood defences fail was provided in 2005, when Hurricane Katrina hit New Orleans. The associated storm surge caused more than 50 breaches of the flood defences built to protect the city, two-thirds of these arising from structural failure. As a result, more than 80 per cent of New Orleans was flooded. The surge, in places in excess of 8 metres in height, widely penetrated 10 kilometres inland, and up to 20 kilometres along river valleys. Shocking as it was for many, to those in the know the swamping of New Orleans came as no surprise. The city is located on the Mississippi Delta, and has been sinking at a rate of about 1 metre a century, due

17. Hurricane Sandy's storm surge—up to 4 metres in places—breached the coastal defences in New Jersey. Causing losses of US$83.9 billion (based on the 2022 Consumer Price Index adjusted cost), Sandy is the fourth most costly hurricane to strike the United States.

mainly to the large-scale abstraction of water from the soft sediment beneath, primarily for irrigation. As a result, Katrina struck a city that—on average—was almost 2 metres below sea level; in places, more than 3 metres. It would not be inaccurate to say that what happened in August 2005 was a disaster waiting to happen. The city continues to subside by up to 5 centimetres a year, and its long-term future, as sea-level rise accelerates, has to be in question.

Other cities built on river deltas, including Alexandria (Egypt), Bangkok (Thailand), Jakarta (Indonesia), Manila (Philippines), and Shanghai (China), have the same problem. Forty per cent of Jakarta is below sea level, and much of the city was sinking at almost 30 centimetres a year, although this has slowed a little due to government action to limit groundwater abstraction. Nonetheless, the coastal flood threat here remains severe, so much so that the government is planning to move the functions of the capital city to the island of Borneo. Many of China's coastally located cities are also subsiding, including the country's most populous urban centre, Shanghai, located on the Yangtze river delta. Shanghai has subsided by almost 2 metres since the 1920s due to groundwater abstraction, but the rate of subsidence has fallen to around 1 centimetre a year on average following the introduction of laws limiting the digging of new wells. The problem has not, however, gone away. China has the largest coastal population on the planet, and coastal cities other than Shanghai are sinking too. By 2050, in the absence of further measures to curtail subsidence, up to 50 million people and US$9.6 *trillion* in assets will be exposed to a one in 100 year coastal flood event.

Future fire and flood

It isn't difficult to appreciate the fact that, as heatwaves and drought conditions grow in intensity and duration, so wildfires will become an ever greater problem. As climate breakdown progresses, large, destructive wildfires are forecast to ignite in previously unaffected countries that are poorly prepared and lack the expertise to fight and control major blazes. Where the fire threat is still small, it will grow. In the UK, for example, drier and hotter summers are forecast to quadruple, to 120, the number of days every year when the risk of fire is classed as very high. In places where major wildfires are endemic and are already having catastrophic consequences, the picture will just get worse. Across the western United States, the frequency of so-called fire weather—the dry, hot, windy conditions that fuel big fires—has increased dramatically over the past half century, and this is a trend that is set to continue, not just in North America but across the planet.

In 2021, wildfires licking at the outskirts of California's South Lake Tahoe (Nevada) resulted in the panicked evacuation of more than 20,000 people. This is a scenario that will become progressively more common as global heating bites harder, and it is only a matter of time before fires begin to roar into city suburbs, menacing tens of thousands of buildings, threatening many lives, and driving mass evacuations. Modifications can be made to buildings to make them more resilient to fire, including the use of non-combustible materials and installing sprinkler systems, but with insurance companies now refusing to insure Californian homes due to the growing wildfire hazard, this is unlikely to happen on a significant scale.

Prospects don't look good on the flood front either. It is projected that a child born in 2020 will face three times as many floods as their grandparents, and it could easily be more. Just a 1 metre rise in sea level will doom the lives and livelihoods of an estimated 250 million people residing in coastal communities, rising to half a billion if the rise touches 2 metres. Meanwhile, the number of people affected by river flooding, worldwide, is forecast to double by 2030 to 132 million. By 2080, close to 350 million people could be affected annually by some form of flooding. Once again, the speed with which extreme weather is taking hold could mean that the true picture may prove to be far worse.

In terms of property damage, the numbers are equally sobering. By 2030, river floods worldwide are expected to cost three times as much as they do now, at more than US$0.5 trillion a year, while coastal flooding costs will rise tenfold to US$177 billion. By 2080, annual losses due to flooding of all types is projected to be in excess of US$4 trillion. In terms of geographical distribution, India, Bangladesh, and Indonesia are set to bear the brunt; by 2030 these three countries accounting for 44 per cent of the world's population annually affected by river flooding, and 58 per cent of the population affected by coastal floods.

To some extent, the increasing incidence of river floods will reflect the growing influence of atmospheric rivers. Modelling suggests that there may be fewer of these, but those that do occur will be longer and wider. Overall, atmospheric river *conditions*, meaning incessant rain accompanied by strong winds, are set to be 50 per cent more common by the end of the century.

In the coastal zone, increased flooding will be facilitated by accelerating rises in sea level combined with bigger and more frequent storm surges, especially in the tropics, where the most powerful hurricanes and typhoons will be more common. The massive increase in all types of flooding expected in coming decades will require a big rethink in terms of mitigation and management. In many instances, flood defences may no longer work, meaning that populations will need to be moved, allowing rivers to do what comes naturally. Protecting low-lying coastlines, in particular, may prove to be a lost cause in many parts of the world, as rising seas threaten to permanently inundate large areas of land. For some small islands, such as those making up the Maldives and the Pacific nation of Kiribati, this may well mean the wholesale decanting of populations to other countries. For small coastal communities all over the world, it will mean rehousing people further inland, and for many of the world's major coastal cities, it will mean the construction of huge and costly defences or the adoption of a policy of managed retreat.

New York City is already building new flood walls and a new drainage system to better handle torrential rainfall, and it is considering a new barrier costing US$120 billion to protect against future storm surges. There are also plans afoot to upgrade the Thames Flood Barrier, which protects London from coastal and river flooding, but these are unlikely to be in place any time soon. Meanwhile, accelerating sea-level rise means that the risk of the Barrier, or any associated defences, being breached or overtopped is increasing every year. Maintaining protection against rising seas would require the construction of a brand new barrier, probably at the mouth of the Thames estuary, which would be a colossal engineering project that would be hugely expensive. In the Netherlands, one-quarter of the land is already below the level of the sea, which is kept out by a complex system of dykes and other works. Well before the end of the century, half the country is likely to be below sea level, so some desperately urgent thinking is required about the country's future viability. This will likely involve making some hard choices between abandoning parts of the country to the North Sea and building yet bigger and higher defences which might very quickly become obsolete.

All in all, it is clear that flooding is going to be an especially pervasive hazard in a hotter world. While measures can be taken to make individual buildings more resilient to flooding, defending at scale against floods and permanent inundation is certain to become increasingly, probably prohibitively, expensive, so that wholesale retreat from areas most affected is likely to become the dominant policy in many parts of the world. This is going to be hugely disruptive, involving, quite possibly, hundreds of millions of people abandoning their homes and moving somewhere safer, but it may well be the only solution. Along with wildfires, storms, droughts, and all the rest of the geophysical mayhem global heating is set to bring or amplify, floods are something we are going to have to do our best to live with.

Chapter 6
Existential threats and systemic shocks

I trailed the idea of the GGE (global geophysical event) in the introductory chapter, and it seems appropriate to bookend this account by investigating in a little more detail these geophysical hazards writ large.

Geophysical threats capable of erasing every last one of us from the face of the planet are somewhat overplayed. When it comes down to it, wiping out eight billion people is not easy, so any event that might be able to do this, of the geophysical variety at least, has to be rather special. Fortunately for us, there is really only one that qualifies: a collision between our world and a comet or asteroid 10 kilometres or more across, on a par with the impactor that finished off the dinosaurs and three-quarters of all species around 66 million years ago. Even then, it might well be the case that pockets of humans survive, most likely in underground bunkers built to provide shelter for a privileged few if or when nuclear Armageddon is unleashed, although whether there would be sufficient numbers of people to ensure our survival as a species in the longer term is a moot point. In any case, impacts on this scale have return periods of a couple of hundred million years, or even more, and logging and tracking the many chunks of rock and ice hurtling around the Solar System has not picked up anything big enough to cause concern, at least in the near to medium term.

Setting aside giant impacts able to see off humanity altogether, there are other geophysical hazards capable of causing systemic shocks to global society and economy. These have the potential to bring our civilization to its knees or, at the very least, to dramatically change the world as we know it. On this list are smaller impacts of 1+ kilometre objects, volcanic super-eruptions, ocean-wide mega-tsunamis, and what I like to call 'strategic' earthquakes—those with the potential to devastate key cities that act as command and control centres for the global economy. Foremost among the threats to our way of life, however, is climate and ecological collapse, which is happening right now.

It came from outer space

Planet Earth is constantly bombarded in the course of its journey around the Sun. Fortunately for us, most of the billions of incoming fragments are tiny and flash into oblivion as soon as they come into contact with our planet's atmospheric shield. Every now and again, however, the Earth collides with something larger. Objects tens of metres across strike with surprising frequency, perhaps every few decades. The last exploded above the surface in 2013 close to the Siberian city of Chelyabinsk, releasing energy equivalent to 30 Hiroshima atomic bombs, shattering windows across a wide area, and injuring 1,500 people (Figure 18). As I write this (June 2023), the National Aeronautics and Space Administration (NASA) asteroid watch dashboard shows me that the next object of around this size will pass our planet today. It will come closer than the Moon, but will still miss us by a good third of a million kilometres. In any case, in terms of causing serious problems for the global population, objects of this size are of minor importance.

What we need to worry about are impactors that are 1 kilometre in diameter, which have the potential to obliterate a country the size of England. Much worse, objects 2 kilometres or more across are capable of wreaking havoc on a global scale. Although barely

18. A spectacular fireball and vapour trail mark the passage of a small asteroid above the Siberian city of Chelyabinsk in February 2013.

equivalent in diameter to twenty soccer pitches laid end to end, such is the prodigious level of kinetic energy (energy of motion) involved in the collision that a 2 kilometre impactor striking land would leave a crater 40 kilometres or so across and cause devastation at a regional or subcontinental scale. The initial blast would be equivalent to the detonation of 100,000 *million* tonnes of TNT, and would erase as many as tens of millions of lives. It would also loft sufficient pulverized debris into the atmosphere to block out the Sun's rays and plunge the Earth into a freezing *cosmic winter* lasting for several years at least. A dramatic fall in photosynthesis rates would ensue, leading to harvest failure on a global scale and worldwide famine that could, according to some studies, wipe out one-quarter of the world's population. The Earth collides with a 1 kilometre object about every 450,000 years on average, and a 2 kilometre body every couple of million years. Such events, then, are rare, but they are certain to occur. Consequently, it would seem sensible to begin to take measures to mitigate the threat; and this is happening.

Sky surveys undertaken over the past few decades have already identified many of the so-called *Near Earth Objects* (NEOs), whose orbits bring them close, at least in astronomical terms, to our world. Of these, just over 100 are comets and the rest asteroids. As of 19 June 2023, 32,211 *Near Earth Asteroids* (NEAs) have been spotted, including more than 10,000 that are 140 metres or more across—big enough to wipe out a city. The total also includes 852 asteroids with diameters of at least 1 kilometre, whose impact would have serious implications for society. A subset of the NEOs incorporates the rather more ominously named *Potentially Hazardous Objects* (PHOs). These are asteroids or local comets that make close approaches to our planet and that have the potential to cause, at the very least, significant regional damage on impact. Most of these are asteroids (PHAs) of which more than 2,300 have been identified, and more are being discovered all the time. None present a serious threat in the short to medium term, and only 145 objects have a collision probability of more than one in 10,000. The greatest known threat is Asteroid 1015955 Bennu, a 0.5 kilometre wide chunk of rock that has about a one in 1,800 chance of hitting us. The highest probability date of impact, however, is between 2178 and 2290, so we have plenty of time to do something about it if the odds on a collision become worryingly short in the intervening period.

Threats presented by objects like Bennu have provided the incentive for studies into possible ways and means of protecting planet Earth from future collisions. These began to bear fruit in September 2022 when the NASA *Double Asteroid Redirection Test* (DART) probe was intentionally crashed into a body known as Dimorphos, a 160 metre wide rock orbiting a bigger asteroid called Didymos. The plan was to evaluate the possibility of diverting an asteroid that, sometime in the future, had our name on it, and it proved to be a great success. The orbital period of Dimorphos was shortened by 32 minutes, which may not sound like much, but it could be critical if accomplished for an object headed our way. Because the Earth takes just seven minutes to travel its diameter in space, 32 minutes could easily mean the difference between certain collision and a near miss; between mega-death and business as usual. Of course, shifting a far bigger object would require much more energy than that provided by the DART probe crash, but the principle has been established, which has to be good news.

While none of the tens of thousands of asteroids and local comets already spotted present a serious threat, this doesn't mean that the risk of collision is zero. There are still likely to be another 15,000 or so NEOs out there with diameters of 140 metres or more, along with dozens larger than 1 kilometre across. One of these could strike our planet completely out of the blue, or we might spot an incoming object with too little time to act. And then there are the so-called *long-period* comets, chunks of rock and ice sometimes 100 kilometres or more across, with orbits that take them on journeys, centuries or millennia long, to the furthest reaches of our Solar System and beyond. We would probably not spot one of these headed our way until it was well within the orbit of Jupiter, just months from collision, and too late to act. Admittedly the chances of this happening are infinitesimally small, but they are not zero.

Atlas shrugged

Meanwhile, back on Earth, there are further geophysical shocks to the system to worry about, and none greater than the next volcanic super-eruption. Mention this term and almost everyone's thoughts turn to Yellowstone in the US state of Wyoming, because this giant volcano has become the poster child for super-volcanoes. Partly—I like to think—this is because of the popular BBC Horizon documentary, *Supervolcanoes* that I helped make in 2000; but, for whatever reason, it has caught the public imagination. Yellowstone is a low-lying volcano, made up of three giant calderas, the biggest up to 80 kilometres across. This colossal crater was formed around two million years ago by the oldest of three titanic eruptions, which ejected ash that fell across

16 states. The other two craters were formed by similarly huge eruptions around 1.2 million and 640,000 years ago.

Continued interest in Yellowstone is fed by the idea that, based on the average return period of these three events, another 'super-eruption' could happen any day. This is not wrong, but we might equally have to wait another 100,000 years, so the threat is being treated with a respect that it perhaps doesn't deserve. This isn't to say that we shouldn't be concerned about a future super-eruption—we should. If the last big Yellowstone blast happened today it would leave the United States and its economy in tatters. The eruption scoured the surrounding countryside with pyroclastic flows, whose gross volumes were sufficient, if spread across the nation, to cover the entire USA to a depth of 8 centimetres, while ash fell as far afield as what is now El Paso (Texas) and Los Angeles (California).

Notwithstanding the regional devastation they invariably cause, the detrimental effect of super-eruptions on the climate is comparable with that arising from a large impact. Furthermore, with an average return period of 50,000 years or so, they happen up to 40 times more frequently than a 2 kilometre asteroid impact. The last super-eruption occurred at Taupo, at the heart of New Zealand's North Island, 26,500 years ago, but the one that attracts the most attention—barring the Yellowstone blasts—is the 74,000-year-old Toba eruption. Located in northern Sumatra, this eruption, possibly the greatest of the last several million years, expelled at least 3,000 cubic kilometres of ash, and excavated a crater (now filled by Lake Toba) 100 kilometres long (Figure 19). Loading of the stratosphere with sulphur gases brought about a so-called *volcanic winter*—severe cooling that lasted for at least six years, and may have resulted in one-third of the planet being covered with snow and ice. Some researchers have even suggested that the blast brought the human race as close as it has ever come to extinction, leaving just a few thousand individuals alive, although this remains a controversial issue. The burning question is,

19. The colossal eruption of Toba, 74,000 years ago, excavated a crater 100 kilometres long and plunged the world into the depths of volcanic winter.

could a super-eruption today have a comparable impact? There is no doubt that the shock to global society and its economy would be massive, and billions could well lose their lives in a global famine triggered by years of bitter cold and consequent failed harvests. A future super-eruption could even be a civilization-destroying event, hurling us back to a dark age existence. It would be very unlikely, however, to wipe out all eight billion of us.

It is not impossible that another super-eruption might happen within the next 100 years, but the probability is very small. Nonetheless, it is worth taking a look at a few potential candidates. Yellowstone will always be in the frame, even though there is a possibility that future outbursts may be on a smaller scale. Two other volcanoes worth keeping an eye on are Laguna del Maule (Chile) and Uturuncu (Bolivia). Both have hosted huge eruptions in the past, and have been swelling rapidly in recent decades in response to a rise of fresh magma. The truth is that we don't know where or when the next super-eruption will breach the surface. What we can be certain of, however, is that it will happen—sometime.

It might come as a surprise, but volcanoes can cause massive problems even without erupting, and this is especially true of giant volcanoes such as those that make up ocean island archipelagoes. As successive eruptions make these volcanoes grow ever larger and taller, they become increasingly unstable until a chunk of the flank falls off in the form of a huge landslide. Imagery of the sea floor surrounding the Hawaiian Islands reveals debris from more than 70 of these huge slides, some with volumes in excess of 1,000 cubic kilometres. Similar landslide deposits are found around other island groups, including the Canary Islands, Cape Verdes, and Azores. The mechanism of volcano lateral collapse, as it is called, does not simply constitute a scientific curiosity, it also has major hazard implications. This is because dumping a huge body of rock into the ocean, at speed, is very effective in generating a tsunami, and not just any tsunami. Deposits of coral and other marine debris are found at elevations of more than 400 metres on the Hawaiian Islands, and shell debris is encountered nearly 190 metres above sea level on Gran Canaria. Both are interpreted as having been emplaced by landslide-driven mega-tsunamis during pre-historic times.

Ocean island volcano collapses will continue to happen at an average rate of around one every 10,000 years so, on a geological, if not human, timescale, another one will likely be along soon. Current concern is focused on the west flank of the Cumbre Vieja volcano on the Canary Island of La Palma. Here, the volcano's west flank became partly detached during an eruption in 1949, and will—at some unknown time in the future—slide into the waters of the Atlantic. Modelling of a worst-case collapse suggests that the resultant tsunami would be hundreds of metres high, and persistent and powerful enough to devastate the entire North Atlantic margin. The scale of destruction would be almost unimaginable, and the cost in US dollars well into the trillions, or even tens of trillions. Without prior evacuation, the number of lives lost would likely be measured in many millions. Inevitably, the shock to the global economy would be unprecedented. It has to be said, however, that while a future Cumbre Vieja collapse will have terrible consequences for the Canary Islands, and possibly the adjacent coast of Africa, there is considerable debate about how destructive the tsunami would be on reaching the North American coast. Some researchers have proposed that at this distance, its energy may have diminished to such a degree as to cause little damage or loss of life. Who proves to be right, only time will tell.

Unlike volcanic eruptions and mega-tsunamis, not even the greatest earthquake is capable of physically impacting the entire planet or even a significant fraction of it. But this doesn't necessarily mean that there are no wider repercussions. When a major earthquake strikes a big city, the consequences are almost invariably horrific, but contained. Where the city is what is described as a global 'command and control centre', however, the ramifications can spread worldwide. Such cities are so deeply invested in the operation of the world economy that any significant disruption can have a hugely destabilizing impact. In recent decades, the traditional command and control centres—London, New York, and Tokyo—have been joined by newcomers Beijing and the San Francisco-San Jose (SFSJ) metropolitan region. Three of these—Tokyo, Beijing, and SFSJ—are under serious threat of a major earthquake, and a fourth—New York—is also not immune to a quake large enough to cause real problems. Only London is seismically safe.

While serious earthquakes striking New York, Beijing, and SFSJ are likely to have significant impacts on the global economic system, it is Tokyo that is the focus of real and immediate concern. This is hardly surprising bearing in mind that, as touched on previously, there is a 70 per cent chance of a major (magnitude 7+) quake striking the Japanese capital in the next 30 years.

A recent local government report forecast that a magnitude 7.3 earthquake, with an epicentre located beneath the south of the city, would take 6,000 lives, destroy or damage almost 200,000 buildings, and leave 4.5 million homeless. Unofficial estimates suggest that such an event would result in economic losses of US$3 *trillion*, almost 70 per cent of Japan's GDP for 2021. And this isn't even a worst-case scenario. A repeat of the magnitude 7.9 Great Kanto Earthquake of 1923 (Figure 20), which destroyed much of the city and took more than 100,000 lives, would bring far greater losses, in terms both of lives and economic cost, the latter having been estimated at US$4.3 trillion. This equates to an entire year's worth of the country's GDP.

It goes without saying that the consequences of such events for the Japanese economy would be shattering. Japan is enormously centralized, and the Tokyo region hosts not only the national government but also the stock market and the greatest concentration of major corporate headquarters in the world.

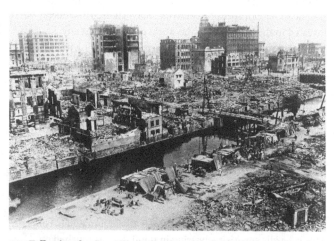

20. Following the Great Kanto Earthquake of 1923, little of Tokyo remained standing after the post-quake conflagrations had raged across the city.

The country is also Asia's second largest economy, and the third largest on the planet, accounting for nearly 4 per cent of global GDP. Despite ever-fluctuating economic circumstances, it remains pretty certain that Japan will still be an economic powerhouse when the big one eventually strikes. In order to rebuild and regenerate it is highly likely that the Japanese will have to disinvest from abroad on a massive scale, dumping government bonds in Europe and the United States, selling foreign assets, and shutting down overseas factories. It is well within the realms of possibility that as country after country finds itself fighting to cope with the resulting financial and economic mayhem, a crash deeper than anything since 1929 or 2008 would soon set in. No one knows how long a post-Tokyo quake depression would last, or just how bad it would be, and much depends on the state of the global economy when the quake hits. Superimposed on a world struggling under the weight of a collapsing climate, its additional impact could result in the world economic order tumbling down like a house of cards.

Climate collapse

It is all very well contemplating geophysical catastrophes to come, but the fact that we do so from the perspective of being at the heart of one makes the exercise somewhat redundant. In reality, ongoing climate breakdown, driven by global heating, represents a whole heap of geophysical hazards, all happening at once, which is why the ramifications are impossible to sidestep.

I am writing this six months ahead of the 28th session of the Conference of Parties (COP28), the latest of the annual climate conferences held every year since 1995, and designed to provide a forum for discussing ways of tackling global heating and reducing greenhouse gas emissions. Between COP1 and COP27 emissions have continued to climb virtually year on year, while the concentration of carbon dioxide in the atmosphere has risen from 360 parts per million to 420 parts per million, so there is a strong argument to be made for the COPs having proved unfit for purpose.

At COP27, convened in Sharm-el-Sheikh (Egypt) in 2022, the focus was very much on keeping the global average temperature rise (since pre-industrial times) below 1.5°C, which, as I touched on in the Preface, is widely regarded as equating to the dangerous climate change 'guardrail'. This is despite the fact that this particular horse has already, to all intents and purposes, bolted. To have any chance of keeping below 1.5°C, emissions need to fall 45 per cent (from 2010 levels) by 2030. While still theoretically possible, in the real world this just isn't going to happen. For the 27th time the elephant in the room was completely disregarded, so that there was no attempt even to discuss cutting production of oil, gas, and coal. To top it all, more than 600 attendees were from fossil fuel corporations, sent along to ensure that nothing was done to jeopardize the sector's interests. With COP28, to be held in the United Arab Emirates petrostate in December 2023, chaired by the head of the national oil corporation, it is hard not to lose all hope of real change happening through these conferences.

There is now a 50 per cent chance that 1.5°C will be surpassed within nine years, and there is no reason not to think that the global average temperature will keep on climbing thereafter. On the basis of current policies, our planet is set to be 2.7°C hotter by 2100, but this is a mean estimate and the true figure could be as high as 3.6°C. If all pledges and promises are fulfilled, which is a huge ask in itself, the rise would still be more than 2°C, quite possibly considerably higher if tipping points are crossed that feed additional heating. Between 2018 and 2022, the global average temperature rise (compared to 1850–1900 levels) has been a shade under 1.3°C. Looking around at how this small rise has boosted extreme weather in just a few years, imagine how bad things are going to be if and when this figure is doubled, or even trebled.

A single terrifying prediction shines a light on where we are headed in the decades to come. According to a 2021 study by the highly respected Chatham House think-tank, by 2050 the world will need half as much food again to feed its still growing population, but climate breakdown will mean that agricultural yields could be down by as much as 30 per cent. If this becomes reality it will be a recipe for widespread starvation, rampant civil strife, and the ripping apart of the fabric of society. Even a smaller yield fall could be catastrophic.

Given such a grim outlook, as early as a few decades hence, it is hard to hold out much hope for the future. But it could be even worse. If all proven fossil fuel reserves are burned, a staggering 3.5 *trillion* tonnes of carbon dioxide will be added to our planet's atmosphere—far more than was emitted between the Industrial Revolution and 2023. This would lead to a double figure hike in the global average temperature that would annihilate ecological systems and make vast areas of the planet uninhabitable for humans. Our civilization would not survive.

This, then—should we fail to curtail fossil fuel extraction and usage—is the end game. Some prefer to deny it, or to apply an optimistic gloss to the truth; others simply pretend it's not happening. It is hard, however, not to agree with legendary animator and story-teller Oliver Postgate, of *Ivor the Engine* and *Bagpuss* fame, who, in his wonderful autobiography *Seeing Things*, observes that 'the real future is what is slowly coming up behind us, but we can't see it clearly because someone has stuck smiley faces over the wing mirrors'.

Further reading

Chapter 1: Hazardous Earth

Jones, Lucy. *The Big Ones: How Natural Disasters Have Shaped Us.*
   London: Icon Books, 2019.

Keller, Edward and DeVecchio, Duane. *Natural Hazards: Earth's
   Processes and Hazards, Disasters and Catastrophes.* Abingdon:
   Routledge, 2019.

Kelman, Ilan. *Disaster by Choice: How Our Actions Turn Natural
   Hazards into Catastrophes.* Oxford: Oxford University Press, 2022.

Knoll, Andrew. *A Brief History of Earth.* New York: Custom
   House, 2021.

Lenton, Tim. *Earth System Science: A Very Short Introduction.*
   Oxford: Oxford University Press, 2016.

McGuire, Bill. *Waking the Giant: How a Changing Climate Triggers
   Earthquakes, Tsunamis and Volcanoes.* Oxford: Oxford University
   Press, 2012.

Molnar, Peter. *Plate Tectonics: A Very Short Introduction.* Oxford:
   Oxford University Press, 2015.

Pelling, Mark. *The Vulnerability of Cities: Natural Disasters and
   Social Resilience.* Abingdon: Routledge, 2003.

Redfern, Martin. *The Earth: A Very Short Introduction.* Oxford:
   Oxford University Press, 2003.

Rothery, David. *Volcanoes, Earthquakes and Tsunamis: A Complete
   Introduction.* London: Teach Yourself, 2015.

Wisner, Ben, Davies, Ian, Cannon, Terry, and Blaikie, Piers. *At Risk:
   Natural Hazards, People's Vulnerability and Disasters.* Abingdon:
   Routledge, 2003.

Zalasiewicz, Jan and Williams, Mark. *The Goldilocks Planet: The 4 Billion Year Story of Earth's Climate*. Oxford: Oxford University Press, 2013.

Chapter 2: Earthquakes and tsunamis

Bryant, Edward. *Tsunami: The Underrated Hazard*. New York: Springer, 2014.

Goff, James and Dudley, Walter. *Tsunami: The World's Greatest Waves*. New York: Oxford University Press, 2021.

Musson, Roger. *The Million Death Quake: The Science of Predicting the Earth's Deadliest Natural Disaster*. London: Palgrave MacMillan, 2012.

Robinson, Andrew and Allen, Daniel. *Earthquake: Nature and Culture*. London: Reaktion Books, 2012.

Thompson, Jerry. *Cascadia's Fault: The Coming Earthquake and Tsunami That Could Devastate North America*. Berkeley: Counterpoint, 2012.

Zeilinga de Boer, Jelle and Sanders, Donald. *Earthquakes in Human History: The Far-reaching Effects of Seismic Disruptions*. Princeton: Princeton University Press, 2021.

Chapter 3: The volcanic menace

Andrews, Robin. *Super Volcanoes: What They Reveal about Earth and the Worlds Beyond*. New York: Norton, 2021.

Branney, Michael. *Volcanoes: A Very Short Introduction*. Oxford: Oxford University Press, 2020.

Heiken, Grant. *Dangerous Neighbours: Volcanoes and Cities*. Cambridge: Cambridge University Press. 2013.

Lockwood, John, Hazlett, Richard, and de la Cruz Reyna, Servando. *Volcanoes: Global Perspectives*. New Jersey: Wiley-Blackwell. 2022.

Lopes, Rosalie. *The Volcano Adventure Guide*. Cambridge: Cambridge University Press, 2005.

Oppenheimer, Clive. *Eruptions That Shook the World*. Cambridge: Cambridge University Press, 2011.

Chapter 4: Storm force

Brayne, Martin. *The Greatest Storm: Britain's Night of Destruction, November 1703*. Cheltenham: The History Press, 2002.

Dolin, Eric. *A Furious Sky: The Five-Hundred-Year History of America's Hurricanes*. New York: Liveright, 2021.

Horne, Jed. *Breach of Faith: Hurricane Katrina and the Near Death of a Great American City*. New York: Times Books, 2008.

Larsen, Erik. *Isaac's Storm: A Man, a Time, and the Deadliest Hurricane in History*. New York: Vintage Books USA, 2000.

Sobel, Adam. *Storm Surge: Hurricane Sandy, Our Changing Climate, and Extreme Weather of the Past and Future*. New York: Harper Wave, 2014.

Chapter 5: Fire and flood

Cook, Margaret. *A River with a City Problem: A History of Brisbane Floods*. Brisbane: University of Queensland Press, 2019.

Johnson, Lizzie. *Paradise: One Town's Struggle to Survive an American Wildfire*. New York: Crown, 2022.

Kelly, Tyler. *Holding Back the River: The Struggle against Nature on America's Waterways*. New York: Simon & Schuster, 2022.

Middleton, Nick. *Rivers: A Very Short Introduction*. Oxford: Oxford University Press, 2012.

Platt, Edward. *The Great Flood: Travels through a Sodden Landscape*. London: Picador, 2020.

Purseglove, Jeremy. *Taming the Flood: Rivers, Wetlands and the Centuries-Old Battle against Flooding*. Glasgow: William Collins, 2017.

Valliant, John. *Fire Weather: A True Story from a Hotter World*. London: Sceptre, 2023.

Chapter 6: Existential threats and systemic shocks

Breining, Greg. *Super Volcano: The Ticking Time Bomb beneath Yellowstone National Park*. Beverly, MA: Voyageur, 2010.

Dillow, Gordon. *Fire in the Sky: Cosmic Collisions, Killer Asteroids, and the Race to Defend Earth*. New York: Scribner, 2020.

Hadfield, Peter. *Sixty Seconds That Will Change the World: The Coming Tokyo Earthquake*. London: Pan, 1995.

Maslin, Mark. *Climate Change: A Very Short Introduction*. Oxford: Oxford University Press, 2021.

McGuire, Bill. *Global Catastrophes: A Very Short Introduction*. Oxford: Oxford University Press, 2014.

McGuire, Bill. *Hothouse Earth: An Inhabitant's Guide*. London: Icon Books, 2022.

Index

For the benefit of digital users, indexed terms that span two pages (e.g., 52–53) may, on occasion, appear on only one of those pages.